Barbuda

This volume explores a range of themes including impacts of climate change, resilience, sustainability, indigeneity, cultural genocide, disaster capitalism, preservation of biodiversity, and environmental degradation. Focusing on the island of Barbuda in the West Indies, it shares critical insights into how climate change is reshaping our world. The book examines how climate has changed in the Caribbean over different spatial and temporal scales and how varying natural and anthropogenic factors have shaped Barbuda's climatic and cultural history. It highlights projections of 21st-century climate change for the Caribbean region and its likely impacts on Barbuda's coastal ecosystems, potable groundwater resources, and heritage. With essays by researchers from the United States, Canada, Caribbean, and Europe, this volume straddles a range of disciplines such as archaeology, anthropology, paleoclimatology, environmental sciences, science education, and Traditional Ecological Knowledge (TEK).

Drawing on interdisciplinary and transdisciplinary approaches that explore the intersection of natural and social systems over the longue durée, the volume will be of interest to scholars, researchers, and students of ethnography, social anthropology, climate action, development studies, public policy, and climate change.

Sophia Perdikaris is Director of Global Integrative Studies (the home of Anthropology, Geography, and Global Studies) and Happold Professor of Anthropology at the University of Nebraska-Lincoln, USA. Her area is environmental archaeology with a specialty in animal bones from archaeological sites. She is interested in people–environment interactions through time and the response of both to big climatic events.

Rebecca Boger is Professor at Brooklyn College, City University of New York (CUNY), USA, and has a background in geospatial technologies, environmental science, and science education. Her research in Barbuda examines socio-ecological resilience, sustainability, environmental/climate change modeling, and community-based mapping.

Critical Climate Studies

Managing Editors: **May Joseph** (US), **Kavita Philip** (Canada)
Commissioning Editors: **El Glasberg** (US), **A. J. James** (India)

The Critical Climate Studies book series is located in the transdisciplinary space that crosscuts the social sciences, humanities, creative writing, environmental studies, and climate science. Scholarship and activism are powerful but often invisible global forces, trapped in the interstices. We seek to draw attention and analysis to such domains. The series welcomes short books that experiment with holistic engagement, critique, and conversation about climate change, broadly conceived. In addition to nuanced academic prose from all disciplines, the series embraces multi-genre writing, experimental ethnographies, creative non-fiction, lyrical sociology, ficto-critical writing, as well as science-humanities collaborations. We encourage contributions that are investigative, immersive, and attentive to the understudied and obscured planetary transformations taking hold as climate change accelerates. Our interests lie in large debates as well as in the understudied regions and microhistories of the world, where the impact of the planet's climate convulsions generate altered experiences and analyses of ontologies, geographies, ecologies, and political economies.

Barbuda
Changing Times, Changing Tides
Edited by Sophia Perdikaris and Rebecca Boger

Aquatopia
Climate Interventions
May Joseph and Sofia Varino

For more information about this series, please visit: https://www.routledge.com/Critical-Climate-Studies/book-series/CCS

Barbuda
Changing Times, Changing Tides

Edited by Sophia Perdikaris and
Rebecca Boger

Routledge
Taylor & Francis Group

LONDON AND NEW YORK

First published 2023
by Routledge
4 Park Square, Milton Park, Abingdon, Oxon OX14 4RN

and by Routledge
605 Third Avenue, New York, NY 10158

Routledge is an imprint of the Taylor & Francis Group, an informa business

© 2023 selection and editorial matter, Sophia Perdikaris; individual chapters, the contributors

The right of Sophia Perdikaris to be identified as the author of the editorial material, and of the authors for their individual chapters, has been asserted in accordance with sections 77 and 78 of the Copyright, Designs and Patents Act 1988.

British Library Cataloguing-in-Publication Data
A catalogue record for this book is available from the British Library

Library of Congress Cataloging-in-Publication Data
A catalog record has been requested for this book

ISBN: 978-1-032-32639-9 (hbk)
ISBN: 978-1-032-39014-7 (pbk)
ISBN: 978-1-003-34799-6 (ebk)

DOI: 10.4324/9781003347996

Typeset in Sabon
by Deanta Global Publishing Services, Chennai, India

This volume is dedicated to the People of Barbuda and all their friends and supporters

In Memoriam Papa Joe and Mrs. Frances Beazer

Contents

Figures

Tables

Contributors

Jennifer Adams is Associate Professor in Education, Creativity, and STEM at the University of Calgary, Canada. Dr. Adams serves as a lead editor for the *Cultural Studies of Science Education* journal. She also serves on the membership committee for the International Society of the Learning Sciences. Dr. Adam's research focuses on urban-place-based and environmental education and informal science learning.

Allison Bain is Professor of Archaeology at Université Laval in Quebec City, Québec, Canada. She specializes in environmental archaeology, paleoentomology, and she is currently working on projects in Québec, Labrador, Iceland, and Barbuda. She also co-directs Université Laval's field school in historical archaeology.

Rebecca Boger is Professor at Brooklyn College, City University of New York (CUNY), USA, and has a background in geospatial technologies, environmental science, and science education. Her research in Barbuda examines socio-ecological resilience, sustainability, environmental/climate change modeling, and community-based mapping. She has been published in several journals including *Quaternary Science Reviews*, *The International Journal of Environmental Sustainability*, and *Environmental Management*.

Michael J. Burn is a Lecturer in Climatology at the Department of Geography and Geology of the University of the West Indies, Jamaica. His research focuses on climatic and environmental change in the tropics, neotropical palaeoclimatology, long-term history of tropical cyclones, and historical biogeography.

Edith Gonzalez is Assistant Professor at the University at Buffalo, SUNY, USA. A historical anthropologist researching the global flow of Traditional Ecological Knowledge, Historical Archaeology of the English-speaking Caribbean, Transatlantic Slavery, 18th-century bioprospecting and human ecodynamics using multiple methodologies, primarily archaeological, ethnographic, and historical. She is a current Fellow of the Eccles Centre for American Studies at the British Library and Research Fellow of the Society of Antiquaries-London.

Noel Hefele received his BFA from Carnegie Mellon and his MA in Arts and Ecology. He paints the interactions between nature and culture in the landscape. He was an artist in residence at the Barbuda Archaeological Research Station. He teaches graduate-level summer Arts and Ecology courses for teachers out of Brooklyn College CUNY, USA.

Jonathan Holmes is Research Director of the Environmental Change Research Centre (ECRC) and Reader in Environmental Change at the University College London, Department of Geography and Professor of Physical Geography, UK. His research is on climate variability in low-latitude regions, rapid climate change events across Europe, and methodological developments in the application of carbonate nonmarine microfossils and microfossil geochemistry to palaeoclimate reconstruction.

Emira Ibrahimpašić is Associate Professor or Practice and Assistant Director of Global Studies in the School of Global Integrative Studies at the University of Nebraska-Lincoln, USA. She is a cultural anthropologist specializing in women's role, identity, ethnicity, and the struggles in a changing political reality in the Balkans and the Caribbean.

Sophia Perdikaris is Director of The School of Global Integrative Studies (the home of Anthropology, Geography, and Global Studies) and Happold Professor of Anthropology at the University of Nebraska-Lincoln, USA. Her area is environmental archaeology with a specialty in animal bones from archaeological sites. She is interested in people–environment interactions through time and the response of both to big climatic events. Since 2005 she has been focusing on the island of Barbuda in the Caribbean, where she explores how heritage work can inform sustainability questions for the future.

Amy Potter is Associate Professor in the Department of Geology and Geography at Georgia Southern University in Savannah, Georgia, USA. Most of her research connects to the larger themes of cultural justice and Black Geographies in the Caribbean and U.S. South. On the island of Barbuda, she has explored a number of topics related to the commons, which include migration, tourism, and agriculture. Her most recent research examines racialized heritage landscapes in the U.S. South. She has published in the *Geographical Review, Journal of Heritage Tourism, Historical Geography, Island Studies Journal,* and *The Southeastern Geographer*. She is also a co-editor of *Social Memory and Heritage Tourism Methodologies* (Routledge).

Russell Leigh Sharman is Assistant Professor of Practice at the University of Arkansas Department of Communication, USA. He is a cultural anthropologist specializing in visual anthropology. He is a filmmaker, writer, director, and anthropologist. He has worked in Hollywood for more than a decade. His accomplishments include studios and production companies like Warner Bros., Disney, and Real FX. His film credits include 'Apartment 4E', which he both wrote and directed, and his book credits include *The Tenants of East Harlem* and *Nightshift NYC*. His ethnographic film work on Barbuda has received recognition awards at the Arkansas Film Festival.

Naomi Sykes FSA is a zooarchaeologist and is currently the Lawrence Professor of Archaeology at the University of Exeter, UK, and researches and teaches on human–animal–environment interactions over the last 10,000 years, exploring how they inform the structure, ideology, and impact of societies, past and present. She is particularly interested in reconstructing the biocultural histories of introduced and locally extinct/ endangered species (e.g., fallow deer, chickens, rabbits, hares, cats, dogs, and wolves). Her approach, which is set out in her 2014 book, *Beastly Questions: Animal Answers to Archaeological Issues*, is to work across the Arts-Science spectrum. She integrates archaeological evidence with data from biomolecular analyses (especially DNA and isotopes) and discussions from anthropology, cultural geography (art) history, and linguistics. Together, these sources of information represent a powerful tool not only for understanding ancient cultures but also for contextualizing modern problems facing humanity.

Acknowledgments

The authors would like to thank the citizens of Barbuda for welcoming them to their island and sharing their knowledge with them. Especially, warm thanks go out to the present and past members of the Barbuda Council and the Barbuda Research Complex, Pastor Nigel Henry of the Barbuda Pentecostal Church along with the principals Ms. Charlene Harris and Mr. John Mussington, teachers, and students from Holy Trinity Elementary School, Sir McChesney George Secondary School, the CUNY REU student participants, the Graduate Center CUNY students from the MALS Sustainability Science and Education track, UNL Education, and all the students that joined us over the years from the institutions represented in our collective affiliations. This work would not have been possible without the help and support of Dr. Reg Murphy and Mr. Calvin Gore, Mr. Shaville Charles, and Mr. Dwight Finch. Special thanks to Dr. Steven Hartman and BIFROST, The Barbuda Research Complex (BRC), SIMARC and Dr. Jay Haviser, Many Strong Voices and Ms Christine Germano, Ms. Louise Thomas and Mr. Mohammid Walbrook. This research was supported by the following: NSF-REU Islands of Change, PSC CUNY-32, PSC CUNY-37, YLACES 2015 & 2019, Ann M. Stack Trust, GEF-Antigua and Barbuda, NSF-OPP Sustainability, Informal Learning and Outreach, the International Community Foundation, and the Groupe de recherche en archéométrie (FRQ-SC) at Université Laval and the Social Science and Humanities Research Council of Canada (820-2010-0189).

Preface

Sophia Perdikaris and Edith Gonzalez

> Who would have thought that the storm blows harder the farther it leaves
> Paradise behind?
>
> – Benedict Anderson (1991: xi)

In early September 2017, Hurricane Irma, a category 5 storm, was bar-
reling toward the small island of Barbuda, its citizens bracing for a direct
hit. Barbuda is part of the dual island, West Indian nation of Antigua and
Barbuda. As the storm approached, the people of Barbuda gathered for
safety in community buildings. Dramatic satellite photos tracking the hurricane's path showcase the moment the island was visible through the eye of
the storm. Then everything went dark.

Two weeks later, a second major storm, Hurricane Jose, was tracking
toward Antigua and Barbuda. After a brief visit to Barbuda, and under the
pretense of new storms following on the heels of Hurricane Irma, the Prime
Minister ordered an evacuation of all Barbudans to its much larger sister
island of Antigua, with no clear plan in place of where or how to house
everyone. People refusing to abandon their homes in Barbuda were forced
to leave at gunpoint by the military. Only a photographer and a guide
remained on the island. All other Barbudans were brought to the larger
island of Antigua to shelter for the duration. Images of Barbudans leaving
what was left of their shattered homes, being towed in a makeshift ferry to
an uncertain future in Antigua, flooded social media. For the first time in
over three centuries, Barbuda was without her people.

As shock and confusion set in among the displaced Barbudan population,
the central Antiguan government's response to the disaster was to impose
a travel ban between Antigua and Barbuda. However, the ban was applied
unequally to Barbudan citizens, corporate entities, and government interests.
The Antiguan government used the chaos engendered by the depopulation
of Barbuda, suspension of 'on-island' Barbudan government proceedings,
and the dispersal of Barbudans as refugees to overturn the Barbuda Land
Act of 2007, which codified a tradition of communal land use. The disman-
tling of the Barbudan policy resulted in a land grab action that successfully
destroyed a system of communal ownership on Barbuda that had been in

existence since the abolition of slavery and reconfigured it into one built on private property and corporate greed.

This volume was conceived to highlight the research which has been done in partnership and collaboration with the people of Barbuda. The partnerships among the members of this research team and the Barbudan people (including but not limited to government officials, guest house proprietors, scientists, shop owners, teachers, school children, professors, elders, farmers, artists, fishermen, hunters, and clergy) were intentional and created thoughtfully by all involved. The original goal for this research was, and continues to be, to present research of relevance guided by island voices and priorities, research that has acknowledged and moved beyond the colonial origins of the academic disciplines represented here. The publication of this work remains imperative as a means of making the research accessible to the people of Barbuda. The research serves as a foundation to enact a Barbudan vision for the future – a vision driven from within. Barbuda has sought solutions to many pre-Irma challenges by using research to inform decision-making. That is why much of the following research focuses on Barbuda pre-Irma; data was gathered for and with the people of Barbuda to address issues, such as an ongoing 'brain drain' and disconnect between youth and elders, environmental sustainability, food security, reliance on imported goods, and threatened cultural heritage.

Barbuda, as we know it, was born in the age of European expansion and during the rise of capitalism in the Caribbean, but who are the people of Barbuda? Barbuda was leased from the English Crown in 1691 by the Codrington family – although the family never settled there, having several estates in Antigua and an ancestral estate in Gloucestershire England (BBC News, 2018). Instead, a group of approximately 100 enslaved people were settled on Barbuda to clear the land and see what agriculture could be practiced there (The Codrington Papers, Barbuda Slave Lists). Over the following 160 years, the culture of Barbuda developed as separate and distinct from Antigua and the surrounding Caribbean islands, with significant influence from the original inhabitants whose African cultural groups are unknown, and the estate manager set to enforce British rule. During the Codrington reign over Barbuda, there were never more than five British settlers on the island, even when the enslaved population grew to over 500 inhabitants. Barbuda never had the revolving door of additional migrants, either voluntary or involuntary, common to sugar production and the Antiguan estates. The crops and livestock were for provisioning other estates and islands. The population of Barbudans then and now are descendent of the original group settled there (The Codrington Papers, Barbuda Slave Lists 1705–1835), allowing for cultural continuity. Generally, in the Anglo-Caribbean, African cultural practices, religious observance, and languages were forbidden on the sugar estates, such as Antigua. Glissant reminds us that because African cultural origins in the Caribbean are, for most, unknown, the Caribbean 'landscape is its own monument: its meaning can only be traced on the underside. It is all history' (Glissant 1992: 11).

The rationale of Barbudans as Indigenous was brought before the Parliamentary Committee on Slavery just prior to the end of slavery within the English empire in 1834 by Christopher Bethel Codrington, the leaseholder of Barbuda and owner of all goods and chattel thereon. Codrington made the claim that the enslaved people of Barbuda were Indigenous and could not be removed from his estate there as a condition of their emancipation. He claimed it would be inhumane that the people were inextricably tied to the land and could not be separated from it without severe consequences to their health. He said that he would rather give the people of Barbuda the use of his land, structures, tools, and transport at no cost than risk their well-being by sending them to another island. While this might indicate ethical/moral conduct, in correspondence from 1829 to 1835 with Robert Jarritt, his estate agent in Barbuda, Codrington stated that he did not believe the experiment of freedom would succeed, so rather than transport the population of Barbuda elsewhere, he chose to immediately emancipate everyone on the island and accept the per-capita compensation from the English government. He appeared to believe that when (not if) emancipation failed, the people of Barbuda would have kept his agricultural enterprise viable and easily be recaptured and enslaved if they remained in situ until the legislation was repealed. Instead, emancipation continued and the case for Barbudan indigeneity remained 'on the books' so to speak. It is important to note that he did not make this same argument of indigeneity for the people enslaved on his Antiguan estates.

In keeping with the guidelines set by Indigenous peoples for the United Nations Forum on Indigenous Peoples, we seek to identify Barbudans in the past as Indigenous, and their own self-identification speak for the present and future, using the following criteria:

- Self-identification as Indigenous peoples at the individual level and accepted by the community as their members
- Historical continuity with pre-colonial and/or pre-settler societies
- Strong link to territories and surrounding natural resources
- Distinct social, economic, or political systems
- Distinct language, culture, and beliefs
- Form non-dominant groups of society
- Resolve to maintain and reproduce their ancestral environments and systems as distinctive peoples and communities

Of special interest in this list are the strong links to territories and resolve to maintain and reproduce ancestral environments, which are cornerstones of Barbudan culture (United Nations Development Group 2009). Edouard Glissant has stated that on other Caribbean islands 'people did not relate even a mythological chronology of this land to their knowledge of this country, and so nature and culture have not formed a dialectical whole that informs a people's consciousness' because of a historical system of forced labor (Glissant 1992: 63). However, the traditional use of land as communal property held in trust for the people of Barbuda, by the people of Barbuda, was a unique system of governance that allowed for an intimacy with nature in Barbudan consciousness. While Glissant's assertion holds true in Antigua, Barbuda's culture and history are permeated by their ties to the land.

Although the Codrington family returned their lease of Barbuda to the English Crown in the 1860s, oral tradition held that the island was bequeathed to all inhabitants by the Codringtons (Russell and McIntire 1966: 2). Antigua and Barbuda became a constitutional monarchy with a British-style parliamentary system of government and formally independent from Britain in 1981. In the late 1980s, only Barbuda had local self-government and continued its communal land system; the other localities of Antigua and Barbuda fell under the authority of the Ministry of Home Affairs (Meditz and Hanratty 1987). In 2007, this communal system was entered into law by the Barbuda Land Act. This act effectively prevented the development of Barbuda by outsiders and required community approval for projects seeking to exploit the natural resources for tourism. The 2007 Land Act was repealed in 2018 (Barbuda Land (Amendment) Act 2017 No. 41 of 2017).

Naomi Klein's 2007 book *Shock Doctrine: The Rise of Disaster Capitalism* takes a deep dive into the economic policies developed by Milton Friedman – an American economist who advised world leaders who advocated for a completely unfettered capitalist system. Klein provides examples of these policies adopted by the US military operations in Iraq in which capitalist entities descend upon a community in the midst of a war zone in order to exploit economic opportunities before the local people can recover from the 'shock'. These policies were used on American soil in the aftermath of Hurricane Katrina in New Orleans.

The cases of Hurricane Katrina and Hurricane Irma are strikingly similar – both were used to push forward policies that served the private interest and had been circulating for years without success. These policies were hotly contested by local people under normal circumstances, but in the suspension of regular localized government operations, policies supported by the majority of local residents were replaced with policies benefiting a select few outside interests. According to Klein, Friedman's economic strategy advocated for governments and corporate entities to use natural or human-made disasters to 'start over with a clean slate' and introduce ideas and policies which could not gain traction in normal times (Klein 2007: 8). In the case of New Orleans, plans to privatize local schools in the poorest wards and close public housing to make room for developers had been repeatedly shut down by local citizens until Hurricane Katrina razed the city. As Klein notes, 'most people who survive a devastating disaster want the opposite of a clean slate: they want to salvage whatever they can and

begin repairing what was not destroyed; they want to reaffirm their related-
ness to the places that formed them' (Klein 2007: 10).

As Klein eloquently states, disaster capitalists have no interest in repairing what was (Klein 2007: 10). In New Orleans, the St. Bernard Housing
Project remained boarded up for two years post-Katrina – forcing residents
to find other housing. As long as people lived there developers could not
make any changes, but the moment the residents were evacuated developers
could use the disaster as a way of 'cleansing the neighborhood' of the poor
black people in it. Charter schools replacing public schools across the poorest districts of New Orleans post-Katrina are just another example of sweep-
ing capitalist policies implemented under disaster response and packaged as
'Hurricane relief.' These policies eroded labor laws, gave away corporate
tax breaks, and gave school vouchers to corporate-run charter schools. All
of this took its toll on neighborhoods and community infrastructure and did
nothing to help the most vulnerable residents (Klein 2007: 6). This pattern
has been replicated in Barbuda, with the storm opening the door to abolish-
ing a resilient traditional community.

Immediately following Hurricane Irma, construction crews with heavy
machinery infiltrated the island and started bulldozing an area for a new airport that would be approximately half the size of LaGuardia International
Airport in New York City. This drastic action was inexplicable, given the
fact that the existing runway of Barbuda's small airport, designed to serve
the 1,800 inhabitants, suffered almost no damage and was completely func-
tioning. The bulldozing of this large tract of land was being done without
the knowledge or consent of Barbudans, or their local government, who
had all been evacuated and were refugees – either living in locker rooms at
the Sir Vivian Richards Cricket Stadium in Antigua or spread out across
Antigua, staying in shelters or with family and friends.

Excavation for the airport continued seven days a week, with stadium
lights placed at the site so work could go on throughout the night. Yet at
the same time, almost nothing was done to repair the damage to peoples'
homes, places of work, schools, the hospital, and roads. The outsized airport
was being built over destroyed cultural heritage sites while Barbudans' traditional land ownership system was quietly being privatized, allowing developers to exploit natural resources unchecked by the local governance. As
you will see in the following chapters, Barbuda was a place with a wealth of
cultural heritage and the ability to give insights into people–environmental
interactions and human response to big weather through time. The research
completed pre-Irma and events post-Irma show in stark contrast the ability of a community to shape its own future based on rigorous research and
culturally sound, long-term sustainable practices.

Infrastructure around the world is falling into neglect, and the pressure
of big weather is knocking out power grids, roads, and bridges, which in
turn are being left to rot. Delay or refusal to repair core services on desirable

real estate is a tactic for development and personal profit. New Orleans was one canary in the coal mine. Even though it is a major mainland US city, its most vulnerable populations and neighborhoods with deep roots in American history were not protected, neither before the storm nor after the levees broke.

> Not so long ago, disasters were periods of social leveling, rare moments when communities put divisions aside and pulled together. Increasingly disasters are the opposite: they provide windows into a cruel and ruthlessly divided future in which money and race buy survival.
>
> (Gould and Lewis 2018)

Barbuda's fate hangs in the balance because of its size and isolation. Who will notice if one more small Indigenous community disappears? In the words of Elizabeth Deloughrey, 'both cultural and biogeographic studies of islands are caught between discourses of migration and island isolation,' and we are encouraged to disrupt 'narratives that emphasize bounded cultures and material practices' (2004: 300). We can no longer continue to consider places like Barbuda bound to a tiny island. This thinking makes us all unsafe.

References

Anderson, B., 1991. *Imagined communities*. 2nd edition. Verso, New York.

Antigua and Barbuda Profile Timeline. BBC News - Latin America, 12 March 2018. https://www.bbc.com/news/world-latin-america-18707512.

Barbuda Land Act, 2017. No. 23 of 2007. Antigua and Barbuda.

Barbuda Land (Amendment) Act 2017 No. 41 of 2017. Published in the Official Gazette Vol. XXXVIII No. 8. January 22, 2018. Government printing Office Antigua and Barbuda.

The British Library, Eccles Centre for American Studies. The codrington papers: Correspondence relating to the Codrington family estates in Antigua and Barbuda. Western Manuscripts, RP 2616.

Deloughrey, E., 2004. Island ecologies and Caribbean literatures. *Tijdschrift voor Economische en Sociale Geografie*, 95(3), pp.298–310.

Glissant, E., 1992. *Caribbean discourse: Selected essays*. Translated by M. Dash. Caraf Books, Charlottesville.

Gould, K. and Lewis, T., 2018. Green gentrification and disaster capitalism in barbuda. NACLA Report on the Americas, Volume 50, – Issue 2: *Eye of the Storm: Capitalism, Colonialism, and Climate Change in the Caribbean*, pp.148–153, 8 June 2018. https://www.tandfonline.com/doi/abs/10.1080/10714839.2018.1479466?journalCode=rnac20

Guidelines on Indigenous Peoples' Issues, United Nations Development Group. 2009, pp.8–10. https://www.un.org/esa/socdev/unpfii/documents/UNDG_guidelines_EN.pdf

Jarritt, R. Correspondence with C.B. Codrington 1829–1835, Oxford, Bodleian Libraries, *Letters concerning the Codrington Family Estates in the West Indies*, MS. 15096, fol. 1, pp.29–62.

Klein, N., 2007. *Shock Doctrine: The rise of disaster capitalism*. Picador, New York.

Meditz, S. and Hanratty, D., 1987. Caribbean Islands: A country study. GPO for the Library of Congress. http://countrystudies.us/caribbean-islands/98.htm

Russel, R. and McIntire, W., 1966. *Barbuda Reconnaissance*. Louisiana State University Press, Baton Rouge.

Introduction

Sophia Perdikaris and Rebecca Boger

There is no world, there are only islands.

– Jacques Derrida (2011: 9)

The notion that the sea level will rise slowly, separating landforms and transforming the globe into islands, is extensively explored through an ecocritical lens on resilience by Jonathan Pugh and David Chandler in their book *Anthropocene Islands: Entangled Worlds* (Pugh and Chandler 2021). The Anthropocene has brought about many challenges that put the emphasis on planetary change resulting from human action. The unfortunate implication of such a statement describing our epoch is that the ones that contribute the least are the ones that are affected the most. This book was written to describe the transition on the island of Barbuda and how an island on the periphery is contributing to our dialogue of place, people, environment, and economy when the power of the people is taken over by big conglomerates, central government, and outside interests. As Nakashima et al. (2012: 11) discuss: 'Small Island societies have lived for generations with considerable and often sudden environmental change. The traditional knowledge and related practice with which small island societies have adapted to such change are of global relevance.'

De Souza frames the importance of understanding the relationship between islands, their environment, and people:

As a home to important flora and fauna, with rich cultural roots and heritage, island communities are often characterized by their deep social ties with the natural environment. However, due to environmental degradation, impacts from climate change including slow (e.g. sea level rise) and sudden (e.g. hurricanes) onset events and the associated changes to livelihood structures and opportunities, islands throughout the world face increasing threats. In order to understand and appropriately address livelihood risks in these communities and to identify opportunities for resilience-building, there is an urgent need to shed light on the

DOI: 10.4324/9781003347996-1

historical and cultural context of island societies and ecosystems. These approaches should build upon local and traditional knowledge and be grounded in established practices developed by island communities over centuries which continue to be heavily impacted by current political and economic trends.

(De Souza et al, 2015: 3)

This volume has been built on pledges like these. It is the science community dialogue that values traditional ecological knowledge (TEK) and contributes science of relevance for a more resilient and sustainable future. The contributions in this book are a tribute to islandness. Barbuda at its core and as the 'canary in the coal mine' is an example of recent practices gone wrong and an example of what can be a global affliction if those practices continue. Can redirected agency slow down the destructive process of capitalist takeovers? What might be alternatives that can inform and redirect actions that instead of contributing to destruction, embrace a more ethical and mindful climate mitigation strategy?

Naomi Klein discusses and coins the term 'disaster capitalism' in her book *Shock Doctrine: The Rise of Disaster Capitalism* (2007) when big events, triggered in this case by anthropogenically induced big weather, not only devastated the local environment but opened the door for colonial takeovers, under the pretense of help, progress, and employment. There has been extensive literature on islands and islanders and the resilience of island peoples (Watts 2018: 198, 75–76), where they respond to critical problem-solving from 'a shared experience of making practical ad hoc solutions'. With disaster capitalism, the ability of islanders to respond is suffocated and stripped away. The deliberate deafness to reason in search of short-lived agendas driven by temporary economic capital gain is not only criminal for Barbuda but also for the global community and the ethics of care and needed advocacy that focuses on the well-being of small islands in climate mitigation.

This monograph is composed of seven chapters presenting a multidisciplinary approach to understanding the long durée (long-time perspective) that has brought Barbuda to experience the full effect of the Anthropocene. The authors have chosen to focus on collaborations with the people of Barbuda in a dialogue of traditional ecological knowledge, ecological consciousness, and the lived experience of climate disasters. As this research was conducted, the Barbudan canary was just entering the mine – but now it is dying. This work captures moments in time for the people and environment of Barbuda pre-Irma, which marks the beginning of disaster capitalism under the guise of benevolent aid and economic growth.

Columbus sent in motion the colonization of the Caribbean, resulting in the hostile takeover of land, degradation of the environment, and subjugation of people forced to labor in exploiting their own resources. The modern economy of the western world has its roots in the emergence of capitalism from 18th-century Caribbean production, with a complex history of forced migration, controlled by European colonizers competing for global markets for cod, salt, spices, tea, slaves, and sugar (Mintz 1985, James 1992, Perdikaris 1999). This era is defined by economic revolution – the drastic increase in economic growth which involves 'the unceasing ascension of the ideology of national economic development as the primordial collective task' (Wallerstein 1988: 4). As Elizabeth Deloughrey points out,

> We could conclude that time/space compression, perceived as intrinsic to the globalized, postmodern era, was long familiar to the Caribbean due to centuries of fragmentation and reassembly in the landscape, economy, and culture. To pursue this logic, tropical islands, far from representing the remote and archaic past, embody the earliest structures of capitalist modernity as well as its contemporary global inequities.
> (Deloughrey 2004: 308)

In the 21st century, little has changed except in the speed and scale of exploitation. Under the pretense of economic growth for local people – envisioned by some as the creation of jobs to replicate a Western-centric ideal of success – the new resource ripe for exploitation is tourism. Similar to colonial acquisition in the past, current land grabs for the development of hotels, airports, and other tourist luxury estates have recreated the power imbalance between outsiders and local people by placing the latter into servitude as day laborers, cleaning staff, and restaurant workers.

Islanders have always been a community of learners because their visceral experiences of responding and coexisting with an environment have made them experts in the day-to-day management of their landscapes. Barbudans have always responded to big weather through community, but because of the effects of the Anthropocene, things are changing faster than ever before. Partnerships with scientists and researchers can expand the local knowledge pool by adding global data so that TEK is kept up to date and resilient enough to meet the here and now. As Rubow (2018: 38) writes, with regard to how big weather events are experienced on islands:

> It may be possible to hold a 'modern' or 'global' perspective on things on a fine clear day, and at a distance to see a cyclone as a discrete weather-object. But when the loud howling noises, the invading waters and crushing boulders enter one's house, the hybrid mess of things and humans is impossible to overlook.

When Hurricane Irma hit the island on September 6, 2017, everything changed for the people of Barbuda. The damage to both built environments and natural environments was extensive. Before the storm, Barbuda was a little-known place to most people around the world. It was thrown into international news in the aftermath of the storm. It is a small island of about 161 km² with a pre-Irma population of about 1,800 people (Boger and Perdikaris 2019). Outsiders might ask, 'Why should we care about what happens to Barbuda? There are so many other problems elsewhere that need our attention'. As fellow humans, we can empathize with the loss of home and property. Yet there are other reasons we, in other parts of the world, should be aware of what is happening to Barbuda. It is a canary in our shared global coal mine, and one of the many victims of disaster capitalism brought on by an extreme event triggered by climate change. Unfortunately, despite international attention, it won't be the last. Barbuda is fraught with challenges due to climate change – sea level rise, increased extreme events of drought, hurricanes, and food security. The island is at a crossroads that is unprecedented in its history, but a cross-road that may soon be on everyone's doorstep (Perdikaris et al. 2021a and b). We have learned that the impacts of climate change and disaster capitalism are powerful forces which must be met with a cultural response. Many insights can be gleaned from examining Barbuda that is relevant around the world.

Sustainability and resilience

The overarching goal of this project is to provide opportunities for informed empowerment, which strengthens the capacity of people to build more sustainable and resilient societies and ecosystems. We examine human–environmental interactions over the past 5,000 years to learn how people have dealt with drought and hurricanes and still provide adequate food and water for their existence. All of our research is socially relevant, and much of it seeks solutions to the impacts of climate change through the integration of TEK and science. Education is woven into everything we do through the inclusion of youth, students, and adults.

The final statements of the Planet Under Pressure conference distilled the thoughts of over 6,000 scientists. They state:

> The continued functioning of the Earth system as it has supported the well-being of human civilization in recent centuries is at risk. Without urgent action, we face threats to water, food, biodiversity and other critical resources: these threats risk intensifying economic, ecological and social crises, creating the potential for a humanitarian emergency on a global scale. In one lifetime our increasingly interconnected and interdependent economic, social, cultural and political systems have come to place pressures on the environment that may cause fundamental changes in the Earth system and move us beyond safe natural boundaries.

But the same interconnectedness provides the potential for solutions: new ideas can form and spread quickly, creating the momentum for the major transformation required for the broadest possible impact – a truly sustainable planet (Planet Under Pressure 2012).

The theoretical approach to our work focuses on the long-term interactions of humans with nature (longue durée) and on the principle of studying the past as a means of understanding response strategies that can assist in resilience planning for the future. Addressing complex global problems like sustainability requires a concerted multidisciplinary perspective. As a social science which integrates multiple methodologies, anthropology is well positioned to facilitate interdisciplinary conversations about sustainability.

The heart of our work is sustainability – defined by Brightman and Lewis 'as the process of facilitating conditions for change by building and supporting diversity—ontological, biological, economic and political diversity' (Brightman and Lewis 2017: 2). An important contribution to an understanding of sustainability comes from anthropological discussions focusing on disasters. Whether it be through archaeological evidence of prehistoric and historic disasters or present-day ethnographic studies of migrations due to climate change, the anthropology of disasters plays a significant role in understanding how humanity confronts these challenges or stressors. Qualitative methods used in anthropology offer a much-needed critical look at 'mainstream policies of sustainability and implementation' (Koensler 2016: 46). Ethnographies are becoming one of the only opportunities for scientists and policymakers concerned with climate change to 'hear people's voices and learn about actual lives affected by their policies and projects' and allows us to 'use the flexible approaches to social problem-solving which sustainability problems require' (Hanchett 2016: 13).

Anthropology offers a holistic perspective that allows us to 'address the critical importance of the assumption of an inextricable link between the physical and the social world' (Hastrup K. and F. Hastrup 2017: 2). Several archaeologists have called for the field to reorient itself solely from looking at the past but also toward the future and archaeology can be especially useful when it comes to looking at current environmental problems with such examples as the archaeological study of the desertification of Sahara and the relocation of human settlements to adjacent regions (Pikirayi 2019: 1660) or recent studies by Peruvian and Chilean archaeologists on the understanding of the origins, patterns, and effects of El Niño weather (Dawdy 2009: 141). Archaeological colleagues working in the Mississippi River delta have contributed to our understanding of the river system's natural dynamics – information useful in designing coastal restoration (Dawdy 2009: 141). These authors argue that archaeology has the potential to illuminate 'the long-term damage we are doing to the planet' (Dawdy 2009: 154) and how archaeology makes an important contribution to the study of sustainability from a longue durée perspective. Our team has completed extensive archaeological survey and rescue excavations on both pre-Columbian and historic sites around Barbuda. Our research explores cause, effect, recovery, adaptation, and extinction in people, place, and space at different time periods, at shorter and longer timescales, using the principles of anthropological archaeology and environmental science. Through the longue durée approach, we aim to contribute to the dialogue toward informed resilience to maximize sustainability but to also identify the factors that are diverting and restricting the island and its people from long-term cultural and ecological survival.

Meeting the needs of the present without compromising the needs of future generations (living sustainably) has been described as the greatest challenge of our generation. It will require international, interdisciplinary, and transdisciplinary efforts to adapt, mitigate, and survive rapid global change and the upheavals it wreaks on society. Effective education for sustainability is needed, in what will certainly prove to be a multigenerational effort, to achieve a soft landing for humans and the planet (Adams et al. 2021). It can and must be part of the bedrock of nation-states and the fabric of societies.

Our research has shown that over the last 400 years, Barbudans and the island ecosystems were resilient and that the Barbudan way of life was sustainable. Barbudans successfully navigated the island's marine and terrestrial ecosystems so that most of their food and water needs were met on an ongoing basis (Boger et al. 2014). People did not stress the island's ecosystems beyond ecological thresholds altering or shifting ecosystem functions. Barbudans lived sustainably within the natural cycles of ecological disturbances (Boger et al. 2016). They accepted and expected natural disturbances of extreme events (e.g., hurricanes and droughts) and successfully navigated through these boom and bust periods (Perdikaris and Hejtmanek 2020). These periods could be viewed through the resilience framework as phases of the adaptive cycle, where the social–ecological system was in dynamic equilibrium. After Irma, Barbudan resilience is being tested in ways they haven't seen before, at least as it pertains to the speed and magnitude of the socioeconomic forces at play. Is this a transformative phase in the adaptive cycle? Or, is this leading to a collapse of the social–ecological system?

For our work in Barbuda, sustainability and resilience are two sides of one coin. To be sustainable, sustained resilience is needed, and to be resilient, people need equitable and economically viable livelihoods in healthy ecosystems.

Structure of this book

The first two chapters set the stage for long-term, place-based examination of human–environmental interactions in Barbuda. Chapter 1, 'A Long-Term Perspective of Climate Change in the Caribbean and Its Impacts on the Island of Barbuda', by Drs Michael J. Burn, Rebecca Boger, Jonathan Holmes, and Allison Bain, reviews how climate has changed in Barbuda on different temporal and spatial scales and the complex network of global and regional forcing factors, both natural and anthropogenic, that make the climate of the island what it is today. Until recently, Barbudans were resilient with a sustainable way of life that leveraged the natural ecology of the land; Barbudan culture and identity thrived with its deep roots in the landscape. In Chapter 2, 'Water Use and Availability on Barbuda from the Colonial Times to the Present: An Intersection of Natural and Social Systems', Dr. Rebecca Boger and Dr. Sophia Perdikaris examine the strategies developed by Barbudans in colonial times, to leverage the island's ecology for sustainable lifeways. It remains to be seen whether or not Barbudan culture can weather the current socioeconomic forces that will constrain their natural resources. This chapter presents new data on water quality and quantity in Barbuda and raises questions about how to use the aquifer system to strengthen Barbudan resilience now and in the near future. The rapidly unfolding events in Barbuda are the result of longer-term trends originating in settler colonialism and culminating in one megastorm.

In Chapter 5, 'From the Far Ground to the Near Ground: Barbuda's Shifting Agricultural Practices', Dr. Amy Potter explores Barbuda's agricultural sector, whose increasing reliance on food imports impacts not only island ecology but also local economic sustainability and community resilience through social networks. Dr. Naomi Sykes, in Chapter 4, 'Fallow Deer: The Unprotected Biocultural Heritage of Barbuda', examines the delicate balance of managing and conserving natural resources, such as the fallow deer population on Barbuda, that have become inextricably entwined with Indigenous cultural heritage.

Chapter 3 turns the focus from a long-term perspective of a changing climate and how people have adapted to their environment to the perspectives of youth expressed through the arts. Dr. Jennifer Adams and Noel Hefele present their research, 'Developing Agency and Resilience in the Face of Climate Change: Ways of Knowing, Feeling, and Practicing through Art and Science', which questions, 'how can art and science be integrated in working with youth in ways for them to learn about place, identity and connections to the natural world?' Their work examines ways to create spaces where young people can make salient connections to place and access their creativity because, like scientists, artists have historically addressed compelling societal issues. Noel, as an artist in residence, worked with Barbudan school children to design and create murals. Young Barbudans coined the words for one mural – 'the sea will rise, Barbuda will survive' powerful, strong words full of optimism. Their vision inspires hope in a world that is often stressed and on the verge of chaos.

As Barbuda's first research facility with registered 'not for profit' status under Antigua and Barbuda's companies' legislation, the Barbuda Research Complex (BRC) is committed to the preservation and conservation of heritage and the environment through education initiatives bringing together local stakeholders and interdisciplinary international scholars. Chapter 6, 'Written with Lightning: Filming Barbuda before the Storm', by Russell Leigh Sharman, chronicles the creation of a film documenting the traditional foodways of the people in Barbuda. The narrative that formed over hours of interviews was one of deep conviction over the traditional subsistence autonomy of the island and the growing threat to sustainability from overdevelopment and state intervention, expressed in the own words of Barbuda's people. Through these projects, BRC bridges natural sciences, social sciences, humanities, arts, traditional ecological knowledge (TEK), and citizen science in a dialogue of discovery and innovation, striving to support Barbuda's cultural and environmental integrity and seeking to ensure relevance and ownership of research on Barbuda by Barbudans.

The island embodies what we all face in the 21st century – impacts of climate change, shifting forms of resilience, eroding sustainability, contested indigeneity, disaster capitalism, preservation of biodiversity, and degrading environments. Hurricane Irma's damage, and the covert construction of the airport, initiated the beginning of a series of actions designed to rob Barbudans of their land to impose an unsustainable Disneyfied (Perdikaris et al. 2021b) western agenda that profited the rich and well connected. We come to know this as disaster capitalism, and the case of Barbuda closely aligns with what Naomi Klein (2007) outlines in her book, *The Shock Doctrine*. In the final chapter, 'Disaster Capitalism: Who Has a Right to Control Their Future?', Dr. Emira Ibrahimpašić, Dr. Sophia Perdikaris, and Dr. Rebecca Boger glean insights from examining Barbuda that are relevant around the world. Those claiming to be 'investors and developers' are not vested in the landscape, and their actions disassociate the people from the land. In this case, they disenfranchise Barbudans from Barbudan identity. Perhaps for the first time in Barbudan history, people may not be able to bounce back from Irma or the inevitable future hurricanes. They may not be able to reclaim their land and maintain their relationship with it.

It is important to remember Barbuda is neither static nor monolithic – there are multiple voices and opinions about how change, which some see as progress, moves forward and is implemented across the population. Pre-Irma, Barbuda was more autonomous and consensus about change was reached through village meetings, healthy debate, and democratic voting. This was only possible within the framework of the Barbudan Land Act of 2007. With the repeal of the Act by the central Antiguan government which took over the decision-making process, Barbudan voices have been silenced. Change has become rapid and uniform, going against local expertise and scientific research and falling victim to colonial behavior hiding behind a mask of economic growth. Recent changes to the landscape, such as the contested airport construction project, pay lip service to ecological conservation while destroying UNESCO Ramsar sites and archaeological sites in the process (Boger et al. 2019; Boger and Perdikaris 2019). It is not the intention of the authors, editors, or contributors to this volume to venture a guess at the motives of the developers. Our work is grounded in the factual understanding of human ecological processes. Perhaps we are in a transformative time in history where research needs a paradigm shift to make sense of the world and to discover new ways to solve problems because the old ways are not working (Syme 2008). This does not mean that disciplinary research is not worthy. Quite the contrary, we would argue that all forms of research are more necessary than ever, and that it is through the synthesis of these different forms that we learn more about the world we live in, and explore directions to make environmental and social systems more resilient and sustainable. At the end of the day, the way that the scientific world communicates must also change. Jargon cannot be left to alienate the islands and peoples with which we are working. This volume hopes to share our community dialogue of Barbudans and researchers with everyone.

References

Adams, J., Perdikaris, S. and Boger, R., 2021. Small Island sustainability education: Engaging you in research and education practices for building sustainable futures. In *Handbook on Caribbean Education*, edited by E. Blair and K. Williams. *Section IV: STEM and Caribbean Education Chapter 16*: 315–332. Information Age Publishing, Charlotte, NC.

Boger, R. and Perdikaris, S., 2019. After Irma, disaster capitalism threatens cultural heritage in Barbuda. https://nacla.org/news/2019/02/12/after-irma-disaster-capi talism-threatens-cultural-heritage-barbuda

Boger, R., Perdikaris, S., Potter, A.E. and Adams, J., 2016. Sustainable resilience in Barbuda: Learning from the past and developing strategies for the future. *International Journal of Environmental Sustainability*, 12(4), pp.1–14.

Boger, R., Perdikaris, S., Potter, A.E., Mussington, J., Murphy, R., Thomas, L., Gore, C. and Finch, D., 2014. Water resources and the historic wells of Barbuda: tradition, heritage and hope for a sustainable future. *Island Studies Journal*, 9(2), pp.327–342.

Boger, R., Perdikaris, S. and Rivera-Collazo, I., 2019. Cultural heritage and local ecological knowledge under threat: Two Caribbean examples from Barbuda and Puerto Rico. *Journal of Anthropology and Archaeology*, 7(2), pp.1–14.

Brightman, M. and Lewis, J., 2017. Introduction: the anthropology of sustainability: Beyond development and progress. In *The anthropology of sustainability*, edited by M. Brightman and J. Lewis, pp.1–34. Palgrave Macmillan, New York.

Dawdy, S.L., 2009. Millennial archaeology: Locating the discipline in the age of insecurity. *Archaeological dialogues*, 16(2), p.131.

Deloughrey, E., 2004. Island ecologies and caribbean literatures. *Tijdschrift voor Economische en Sociale Geografie*, 95(3), pp.298–310.

Derrida, J., 2011. *The Beast and the sovereign*, Volume II, Geoffrey Bennington (tr.), University of Chicago Press, Chicago, IL.

De Suza, R. M., Henly-Shepard, S., McNamara, K. and Fernando, N., 2015. Re-framing Island nations as champions of resilience in the face of climate change and disaster risk. UNU-EHS Working Paper Series, No. 17. United Nations University Institute of Environment and Human Security, Bonn. http://collections. unu.edu/eserv/UNU:2856/Reframing_island_nations_WP_No_17_PDF

Gaffney, O. and Broadgate W., 2012. *Planet under pressure: New knowledge towards solutions*. Conference, London. www.planetunderpressure2012.net

Hanchett, S., 2016. Sanitation in Bangladesh: Revolution, evolution, and new challenges. *CLTS Knowledge Hub Learning Paper* 3, Institute of Development Studies, Brighton.

Hastrup, K., and F. Hastrup, eds., 2017. *Waterworld: Anthropology in fluid environments*. Berghahn Books, New York.

James, C.L.R., 1992. *The C.L.R. James Reader*. A. Grimshaw, ed., Blackwell, Oxford.

Klein, N., 2007. *Shock Doctrine: The rise of disaster capitalism*. Picador, New York.

Koensler, A., 2016. Chapter three: The emerging sphere of resonance: "Clandestinely Genuine" food networks and the challenges. In *Envisioning sustainabilities: Towards an anthropology of sustainability*, edited by F. Murphy and P. McDonagh, p.37. Cambridge Scholars Publishing, Newcastle upon Tyne.

Mintz, S., 1985. *Sweetness and power: The place of sugar in modern history*. Viking, New York.

Nakashima, D., McLean, C. G., Thulstrup, H., Castillo, A. M. and Rubis, J., 2012. *Weathering uncertainty: Traditional knowledge for climate change assessment and adaptation*. UNESCO, Paris.

Perdikaris, S., 1999. From chiefly provisioning to commercial fishery: Long-term economic change in Arctic Norway. *World Archaeology*, *30*(3), pp.388–402.

Perdikaris, S., Boger, R., Gonzalez, E., Ibrahimpašić, E., and Adams, J., 2021a. Disrupted identities and forced nomads: A post-disaster legacy of neocolonialism in the island of Barbuda, Lesser Antilles. *Island Studies Journal*, 16(1), pp. 115–134.

Perdikaris, S., Boger, R., and Ibrahimpašić, E., 2021b. Seduction, promises and the disneyfication of Barbuda Post Irma. *Translocal Contemporary Local and Urban Cultures Journal*. No. 5 (un)inhabited spaces. Ana Salgueiro and Nuno Marques (eds). ISSN 2184-1519. Available at http://translocal.cm-funchal.pt/2019/05/02/revista05/.

Perdikaris, S., and Hejtmanek, K.R., 2020. The sea will rise, Barbuda will survive: environment and time consciousness. *Ecocene: Cappadocia Journal of Environmental Humanities*, Special Issue [publication expected December], S. Perdikaris, guest editor, No. 1. 2 (December), pp.93–109.

Pikirayi, I., 2019. Sustainability and archaeology of the future. *Antiquity*, 93, pp.1669–1671.

Pugh, J. and Chandler, D., 2021. *Anthropocene island: Entangled worlds*. University of Westminster Press, London.

Rubow, C., 2018. Woosh-cyclones as culturalnatural whirls: The receptions of climate change in the Cook Islands. In *Pacific climate cultures: Living climate change in oceania*, edited by T. Crook and P. Rudiak-Gould, pp.34–44. DeGruyter Open, Warsaw.

Syme, G.J., 2008. Sustainability in urban water futures. In *Troubled waters: Confronting the water crisis in Australia's cities*, edited by P. Troy. ANU E Press, Canberra.

Wallerstein, I., 1988. The modern world-system as a civilization. Thesis Eleven, No. 20,1988. Sage Journals.

Watts, L., 2018. *Energy at the end of the world: An Orkney Islands Saga*. MIT Press, Cambridge, MA.

1 A long-term perspective of climate change in the Caribbean and its impacts on the island of Barbuda

*Michael J. Burn, Rebecca Boger,
Jonathan Holmes, and Allison Bain*

1.1 Introduction

On September 6, 2017, Hurricane Irma devastated the island of Barbuda, making landfall as Category 5 hurricane at 0545 UTC with catastrophic maximum sustained winds of 178 miles/h (155 kt) and a minimum pressure of 914 mb (Cangialosi et al. 2018). It went on to cause widespread devastation across the Caribbean and the United States, becoming one of the strongest and costliest hurricanes on record in the tropical Atlantic. A tide gauge on Barbuda recorded a peak water level of 2.4 m (7.9 ft) mean higher high water (Cangialosi et al. 2018), indicating a storm surge of at least 2.4 m above ground level for parts of the island. Ninety percent of structures and homes were either damaged or destroyed, including the island's airport and the water supply, and communications were completely cut off. The subsequent threat of Hurricane Jose meant that the entire island was evacuated to Antigua, leaving Barbuda abandoned and uninhabited for the first time in its long and complex settlement history. At the time of writing (January 2021), schools and several shops have reopened and over half of the inhabitants have returned and the fishing season is being revived, and large-scale excavations are underway for tourist developments (Perdikaris et al. 2021).

The physical and socio-cultural devastation that Hurricane Irma left in its wake emphasizes the fragile reciprocal relationships between human settlement on Small Island Developing States (SIDS) and climatic change. It also encourages the inquisitive mind to ponder not only the extent to which island settlement or abandonment was influenced by natural climate variability in the past, but also how anthropogenic climate change may impact society toward the latter half of the 21st century and beyond. In this chapter, we review how climate has changed in Barbuda on different temporal and spatial scales and emphasize the complex network of global and regional forcing factors, both natural and anthropogenic, that make the climate of the island what it is today. It is impossible to understand the past, present, and future climate of Barbuda without placing it in the broader context of

DOI: 10.4324/9781003347996-2

regional (Caribbean-wide) and global climate variability. Thus, our discussions also include comprehensive analyses of climate dynamics on different spatial scales.

Barbuda is a small (161 km²), low-lying limestone island of the Lesser Antilles island arc (Figure 1.1), which separates the tropical Atlantic Ocean from the Caribbean Sea (Brasier and Donahue 1985). Barbuda's maximum

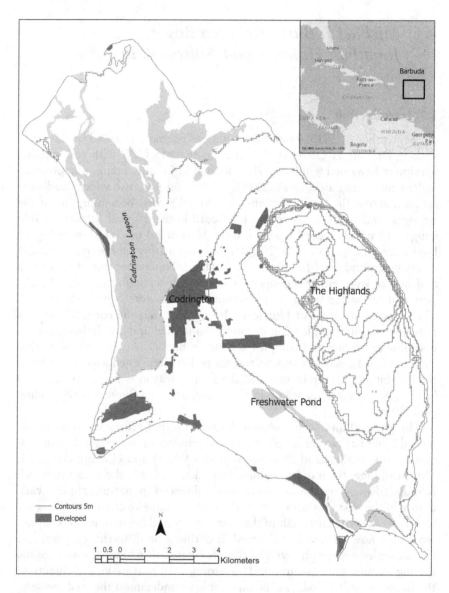

Figure 1.1 Map of Barbuda.

average elevation of ~38 m corresponds to a massive limestone plateau of Pliocene age known as 'The Highlands', which occupies the southeast segment of the island. The plateau is bounded by abandoned sea cliffs to the north and east and by alternating late-Pleistocene consolidated beach ridge and lagoon deposits to the west and south, which form the Codrington Limestone Group (Martin-Kaye 1959). The majority of Barbuda's inhabitants live in the village of Codrington located to the east of the island adjacent to Codrington Lagoon, a semi-enclosed coastal lagoon. Mangrove forests surround the island and are home to the second-largest frigate bird colony in the region. A rare inland mangrove forest comprising extensive stands of *Rhizophora mangle* (red mangrove) fringes Freshwater Pond, one of just a few sources of fresh brackish water on the island. This mangrove ecosystem may be a relic of a more extensive forest that was connected previously to the Caribbean Sea (Stoddart et al. 1973). Fringing coral reefs surround the island in several locations, particularly to the south in White Bay. The island's limestone geology is porous and supports a network of wells across the island, which feed into shallow aquifers with varying ranges of brackish water. These wells were important historically for several purposes including cattle ranching, agriculture, washing, and drinking. Today, they are used less but are still important for watering domestic and wild animals, construction activities, and as sources of potable water for hunters (Boger et al. 2014).

1.2 Contemporary climate dynamics of the Caribbean

The climate of the Caribbean is characterized by a seasonal subtropical maritime climate with average temperatures of around 27°C and a minimum/maximum temperature range of 17–35°C, which remains fairly consistent across the region. The annual precipitation cycle exhibits greater spatial variability than temperature but may be generalized broadly by a dry season from December to April and a wet season from June to November (Gamble and Curtis 2008; Taylor et al. 2002; Curtis 2013). Precipitation is controlled by the interplay between thermodynamic processes associated with Sea Surface Temperature (SST) variability and dynamic processes associated with regional wind shear. These vary in response to the seasonal and interannual migration of the Intertropical Convergence Zone (ITCZ) and associated changes in the North Atlantic Subtropical High (NAH) and the intensity of the Caribbean Low-Level Jet (CLLJ) (Wang 2007; Cook and Vizy 2010; Moron et al. 2016; Figure 1.2), a regional extension of the northeast trade winds. In general, the seasonal rainfall pattern is bimodal, though it exhibits considerable variability across the region. Bimodality is manifested by an early and a late rainfall peak in April–July and August–November, respectively (Chen and Taylor 2002) and is separated by the Mid-Summer Drought (MSD), the magnitude of which decreases along a gradient from Central America in the west through the Greater Antilles (Cuba, Jamaica,

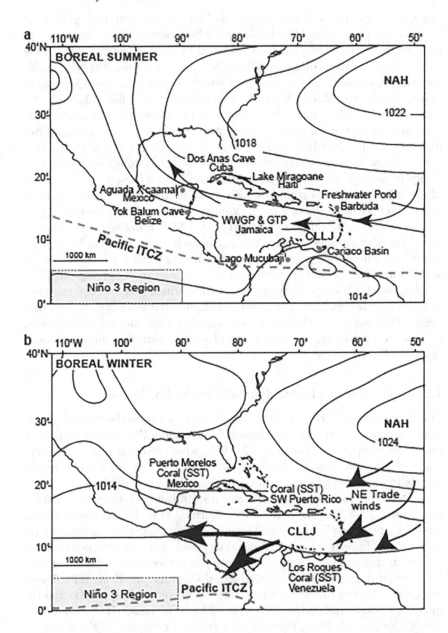

Figure 1.2 Seasonal summer (a) and winter (b) patterns of the Intertropical Convergence Zone (ITCZ), the North Atlantic Subtropical High (NAH), and the Caribbean Low-Level Jet (CLLJ).

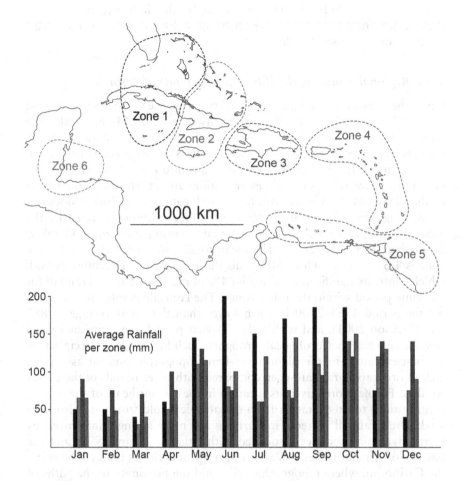

Figure 1.3 Seasonal monthly average precipitation by zone in the Caribbean.

Hispaniola, and Puerto Rico) to the islands of the Lesser Antilles in the east (Figure 1.3; Magaña et al. 1999; Curtis and Gamble 2008; Small et al. 2007; Gamble et al. 2008; Angeles et al. 2010).

During the hurricane season (June–November), precipitation is also associated with pulses of African Easterly Waves, which transport moist and unstable air masses from the African Tropics across the region and are instrumental in the genesis of many tropical cyclones (Landsea 1993). During the winter dry season, lower SSTs and the southward migration of the ITCZ enable regional-scale subsiding air masses associated with the NAH to dominate, in turn suppressing rainfall. Any rainfall that does occur during the boreal winter is mostly localized and associated with the passage of extratropical cold fronts (Moron et al. 2016) that influence the islands of the Greater Antilles most and, given their more distal easterly position, least for those of the Lesser Antilles.

1.2.1 Regional climate variability and the climate of Barbuda

Given the regional precipitation variability across the Caribbean (Gouirand et al. 2012; Magaña et al. 1999), Stephenson et al. (2014) subdivided the region into six distinct climatic zones based, among other parameters, on the seasonal rainfall cycle and the timing of the summer precipitation maximum (Figure 1.3). The island of Barbuda falls into Zone 4, which encapsulates most of the islands of the Lesser Antilles and Puerto Rico. In contrast to the islands of the Greater Antilles, the climatology of Zone 4 does not show a significant bimodal distribution of precipitation separated by the MSD. Instead, it is replaced by a single late-season maximum in October/November (Jury et al. 2007). The mean total annual precipitation for the zone is 1,311 mm of which 80% falls during the May–December period, which contrasts significantly with 55–64% of mean total annual rainfall for the same period within the other zones. The equivalent value for Barbuda for the period 1965–2000 is much lower than the zonal average at 882 mm (Jackson 2001), and within this 35-year period, ten individual years were designated 'meteorological' droughts. Such low levels of precipitation are best explained by the island's low-lying topography and the associated lack of orographic rainfall when compared with other islands of the Lesser Antilles. Furthermore, given its location in the far northeast of the region, contributions to precipitation from extratropical cold fronts are also minimal. Thus, rainfall patterns in Barbuda are more strongly influenced by thermodynamic processes associated with Atlantic summer SSTs and the passage of African Easterly Waves and tropical cyclones than elsewhere in the Caribbean, where topographic relief and the proximity to the paths of extratropical cold fronts complicate the precipitation signal.

On interannual and multidecadal timescales, Barbuda's rainfall is controlled by the relative influence of El Niño Southern Oscillation (ENSO), the Atlantic Meridional Mode (AMM; Kossin and Vimont 2007), the Atlantic Multidecadal Oscillation (AMO; Klotzbach 2011; Winter et al. 2011), and the North Atlantic Oscillation (NAO; Cook and Vizy 2010; Gouirand et al. 2012; Wang et al. 2008; Wang 2007). Interannual changes in SSTs in the eastern equatorial Pacific associated with ENSO dynamics have been shown to be strongest in the NW Caribbean, diminishing somewhat toward the Lesser Antilles (Jury et al. 2007). During a developing El Niño (La Niña) event (September–December), positive (negative) SST anomalies develop in the Pacific, increasing (decreasing) the surface pressure gradient between the NAH and the eastern equatorial Pacific, in turn strengthening (weakening) the CLLJ. Enhanced (reduced) vertical wind shear and increased (decreased) subsidence interfere with (promote) the vertical development of tropical cells, subsequently reducing (increasing) precipitation and tropical cyclone activity and resulting in dry (wet) conditions across much of the Caribbean. Above-normal rainfall in the Caribbean during the boreal spring following an El Niño event (April–May) is common and associated with a warming of the tropical Atlantic (Taylor et al. 2002; Taylor et al. 2011).

The Atlantic Meridional Mode (AMM) is a dynamical mode of the coupled ocean and atmosphere variability in the Atlantic and similarly operates on interannual timescales (Kossin and Vimont 2007). It manifests itself as a cross-equatorial gradient in SST and a shift in the ITCZ toward the warmer hemisphere. A positive (negative) AMM decreases vertical wind shear and surface pressure, in turn shifting the ITCZ northwards (southwards) and creating environmental conditions conducive (unfavorable) to rainfall and tropical cyclone activity. On multidecadal timescales, the AMM is strongly correlated with, and thought to influence, the Atlantic Multidecadal Oscillation (AMO), which is an index of North Atlantic SST variability with alternating warm and cold phases over periods of 65–70 years (Schlesinger and Ramankutty 1994). It has been shown to have a widespread influence on climatic phenomena, including tropical Atlantic hurricane activity (Goldenberg et al. 2001; Klotzbach 2011) and precipitation in Africa (Folland et al. 1986; Knight et al. 2006), the Caribbean (Stephenson et al. 2014), and North America (Enfield et al. 2001), and these oscillations are clearly recorded in annual resolution coral-based SST near Puerto Rico (Kilbourne et al. 2008; Saenger et al. 2009) and across the region (Hetzinger et al. 2008; Vásquez-Bedoya et al. 2012; Tierney et al. 2015).

The NAO affects Caribbean rainfall predominantly during the boreal winter through its influence on the strength and position of the NAH (Wang 2007; Cook and Vizy 2010) and in turn the CLLJ. However, given that its influence is manifested most strongly during the annual precipitation minima of the dry season (particularly so for Barbuda), it is thought to exert less control on total mean annual rainfall than ENSO and the AMM/AMO. Nevertheless, a positive NAO in the boreal winter is characterized by a stronger NAH and CLLJ that combine to cool the ocean surface and advect moisture toward the southwest into the Gulf of Mexico and Central and South America (Martin and Schumacher 2011). The combination of moisture loss and enhanced surface pressure suppresses atmospheric deep convection causing drier conditions in the circum-Caribbean region. In contrast, during a negative NAO phase, the intensities of the NAH and CLLJ diminish, promoting convective rainfall.

1.3 The climate of the past

1.3.1 Holocene sea level rise

One of the principal factors controlling long-term climatic change on Caribbean islands is the orbital forcing of global climate. During the last glacial maximum (ca. 22,000 years BP; unless otherwise states all dates are quoted in calendar years before the present), average Caribbean temperatures were 5–8°C cooler than today and eustatic sea levels 121 ± 5 m lower (Fairbanks 1989). Orbitally forced warming during the Holocene resulted in the melting of the Laurentide and Fennoscandian ice sheets and an increase in eustatic sea levels to their present-day level around 2–3000 years BP. This resulted in significant landscape changes across the region, particularly as it relates to island biogeography and human settlement patterns (Siegel et al. 2015). Indeed, the surface geology of Barbuda is young and has remained relatively stable tectonically throughout the Pleistocene (Brasier and Donahue 1985). Consequently, the island has been shaped by eustatic sea level changes associated with successive glacial cycles. Furthermore, the late Holocene origin of Freshwater Pond, from which a reconstruction of local changes in effective moisture was derived, is thought to be the result of rising eustatic sea level that reached its present-day maximum level ca. 2–3000 years BP (Fairbanks 1989; Burn et al. 2016). The subsequent development of a rainfall-derived freshwater lens, which rests above the underlying saltwater table would have provided a valuable new source of fresh water to settled and/or migrating communities on the island.

1.3.2 Caribbean climate variability during the Holocene

Only a few Caribbean climate records extend back through the Holocene into the last glacial maximum. The most continuous record is that of titanium (Ti) abundance from the Ocean Drilling Program (ODP) Site 1002 sediment archive from the Cariaco Basin, Venezuela (Figure 1.4). Haug et al. (2001) argue that terrigenous erosion of river catchments along the north coast of Venezuela transports sedimentary titanium to the Caribbean Sea, which subsequently sinks and becomes emplaced within the sediments of the Cariaco Basin. The sediment record, therefore, reflects changing rainfall amounts associated with the long-term migration of the ITCZ during the late glacial and Holocene periods (since ~22,000 years BP) where increased (decreased) Ti counts represent higher (lower) rainfall. The timing and sinusoidal characteristics of the Ti record (Figure 1.4) suggest that the orbital precession cycle, which modifies the intensity of the seasonal cycle, is responsible for these long-term changes. Given this interpretation, the late glacial and early Holocene periods were probably characterized by a more southerly position of the ITCZ, resulting in relatively dry conditions across the Caribbean. Regional aridity at this time is further supported by terrestrial paleoclimate reconstructions from Haiti (Hodell et al. 1991), Jamaica (Street-Perrott et al. 1993; Holmes et al. 1995; Holmes 1997), Cuba (Fensterer et al. 2013), and Venezuela (Curtis et al. 2001), suggesting this to be a Caribbean-wide phenomenon.

By the Mid-Holocene (~7000–3000 years BP), the records suggest the mean location of the ITCZ shifted to the north and wetter conditions

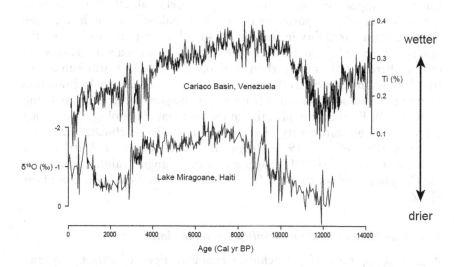

Figure 1.4 Caribbean climate records reconstructed from sediment cores taken in the Cariaco Basin, Venezuela, and Lake Miragoane, Haiti.

prevailed across the region. Such conditions were likely accompanied by changes in the biogeography of the Caribbean, as is well-exemplified in the lowlands of Haiti, where a transition from drought-adapted sclerophyllous vegetation to mesic forest occurred (Higuera-Gundy et al. 1999). Moister conditions were probably also more conducive for human habitation and may indeed explain the initial colonization of the Caribbean ~6000 years BP (Keegan and Hofman 2017; Rousseau et al. 2017; Newsom and Wing 2004; Wilson 2007) and subsequent waves of migration including that of the Saladoid culture ~2500 years BP (Keegan et al. 2013; Keegan and Hofman 2017; Bain et al. 2018; LeFebvre et al. 2019). Since the late Holocene (~3000 years BP to present), both the Venezuelan (Cariaco Basin and Lake Valencia) and Haitian archives suggest progressive drying implicating the precession-driven southward migration of the ITCZ. In contrast, terrestrial paleoclimate reconstructions from Cuba (Fensterer et al. 2012) and Jamaica (Street-Perrott et al. 1993) suggest a trend toward wetter conditions during this period. The apparent disparities between these records may be explained by a transition toward increased regional variability in effective moisture across the region during the late Holocene, a change consistent with the variability observed in the contemporary climatology of the Caribbean (see above).

It is tempting to test the hypothesis that precession-forced and consequently time-transgressive climate variability was responsible for the different waves of human migration across the Caribbean. However, while some suggest that orbitally forced changes in precipitation in the

Caribbean appear time-transgressive (e.g., Siegel et al. 2015), this contention is not yet fully supported by the available paleoclimate archives, given the significant variability in the climate signal among archives during the late Holocene and the significant uncertainties associated with hard-water errors inherent in the radiocarbon chronologies of many Caribbean terrestrial lake sediment records (Curtis et al. 2001) and recent re-examination of multiple radiocarbon dates from archaeological archives (Napolitano et al. 2019). Further, complexities in the regional climatology and strong evidence of non-stationarity of the different modes of climate variability over time (Rodbell et al. 1999; Gray et al. 2004; Metcalfe et al. 2010; Fritz et al. 2011; Burn et al. 2016) suggest that the changing rainfall patterns in the Caribbean are not simply a function of the latitudinal migration of the ITCZ.

1.3.3 Caribbean climate variability during the late Holocene

A growing number of paleoclimate reconstructions of climatic dynamics during the last millennium now exist for the Caribbean region. These include SST reconstructions dating back to the mid-16th century (Kilbourne et al. 2008; Hetzinger et al. 2008; Vásquez-Bedoya et al. 2012; Tierney et al. 2015), lake sediment records of changing effective moisture (e.g., Hodell et al. 2005; Lane et al. 2011; Burn and Palmer 2014; Burn et al. 2016), high-resolution speleothem (e.g., Kennett et al. 2012; Fensterer et al. 2012), and tree-ring records (Trouet et al. 2016) and documentary archives of climate phenomena (García-Herrera et al. 2005; Chenoweth 2007; Chenoweth and Divine 2008, 2014; Berland et al. 2013; Berland and Endfield 2018). Each of these reconstructions provides a snapshot of climate variability at a specific location at different temporal resolutions. Here, we focus on climate reconstructions relevant to specific archaeological periods at both the regional and local scales.

1.3.3.1 Climate variability during the colonial period

The colonial period in the Caribbean occurred at a time when average temperatures in the circum-North Atlantic region dropped significantly during the period known as the Little Ice Age (LIA). Paleoclimate and historical records reveal significant reductions in North Atlantic atmospheric temperatures (Mann et al. 2009; Jones et al. 2009), SSTs (Keigwin 1996; Marchitto and DeMenocal 2003), as well as increases in Arctic sea-ice extent (Broecker 2000; Vare et al. 2009) and alpine glacial advances (Holzhauser et al. 2005). These climatic changes are thought to be attributed to reduced radiative forcing caused by a combination of lower solar activity resulting from four grand solar minima (Wolf, Spörer, Maunder, and Dalton) and increased volcanic activity (Wanner et al. 2008; Jones et al. 2009; Mann et al. 2009; Bindoff et al. 2013).

The paleoclimate evidence from across the Caribbean confirming the footprint of the LIA is reasonably abundant (Haug et al. 2001; Hodell et al. 2005; Lane et al. 2011; Burn and Palmer 2014). Most records suggest the LIA manifested itself as a cooler and drier period punctuated by region-wide drought events, which would have resulted in depleted water resources and acted as a stressor to established cultures across the region. Individual paleoceanographic reconstructions suggest that Caribbean SSTs may have decreased by ca. 2–3°C during the LIA (Winter et al. 2000; Watanabe et al. 2001; Nyberg et al. 2002; Haase-Schramm et al. 2003; Black et al. 2007; Saenger et al. 2009), which would have suppressed rainfall by inhibiting both the northward extension of the ITCZ and atmospheric convection. However, a more statistically robust composite record of coral-based SST reconstructions for the tropical Atlantic and Caribbean suggests the regional drop in mean SSTs decreased by just 0.6°C during the LIA (Tierney et al. 2015). Nevertheless, the impact of the latter estimate would have similarly reduced the amount of available thermal energy, which in turn lowered evaporation rates and resulted in significant reductions in precipitation. There would also have been a significant reduction in rainfall associated with suppressed tropical cyclone activity, the genesis of which requires SSTs in excess of 26°C (Gray 1968).

A suite of terrestrial paleoclimatic archives supports the coral-based pale-oceanographic reconstructions that indicate lower SSTs resulted in drier conditions during the LIA (Haug et al. 2001; Hodell et al. 2005; Metcalfe et al. 2010; Lane et al. 2011; Fensterer et al. 2012). The titanium record from the Cariaco Basin (Haug et al. 2001) and oxygen isotope measurements from a lake sediment record in Aguada X'caamal (Hodell et al. 2005) on the Yucatan Peninsula in Mexico suggest dry periods punctuated the circum-Caribbean region between 1450 and 1800 CE. Further support for a coherent regional drought response to low natural radiative forcing during the LIA may be garnered from a 1500-year reconstruction of glacial advance from a proglacial lake sediment record recovered from Lago Mucubaji in the Venezuelan Andes (Polissar et al. 2006). The magnetic susceptibility record from this site differentiates between sedimentation of clastic minerogenic material and authigenic sediment production. Higher levels of clastic sedimentation during the Wolf, Spörer, Maunder, and Dalton grand solar minima are interpreted as periods of glacial advance. Ice accumulation occurred because of the increased orographic rainfall associated with a stronger CLLJ, which enhanced the flux of advected moisture to the Venezuelan Andes. Together, these sites support inferences drawn from other paleoclimatic archives within the circum-Caribbean region (Hodell et al. 2005; Metcalfe et al. 2010; Lane et al. 2011) that suggest a strong influence of natural radiative forcing on regional drought patterns over the last millennium.

In contrast, recent paleoclimate reconstructions from Belize (Kennett et al. 2012), Cuba (Fensterer et al. 2013), Jamaica, and Barbuda (Burn and

Palmer 2014; Burn et al. 2016) suggest that long-term changes in effective precipitation during the LIA were much more variable, temporally and spatially, in turn indicating that the coupled ocean–atmosphere climate modes ENSO and AMO may override, or are at least superimposed upon, the influence of natural radiative forcing. Kennett et al. (2012) present results from a high-resolution speleothem record from Belize and argue that precipitation variability was driven by a combination of the long-term migration of the ITCZ and ENSO dynamics with drier (wetter) periods associated with El Niño-like (La Niña-like) conditions. They draw on contemporary climatology to argue that during an El Niño event, during which positive SST anomalies dominate the tropical Pacific, a decrease in surface pressure associated with the Pacific ITCZ increases the North Atlantic-equatorial Pacific pressure gradient, in turn resulting in a stronger CLLJ, enhanced wind shear and drier conditions across the Caribbean. Moreover, Fensterer et al. (2012) present a high-resolution oxygen isotope record from a speleothem in western Cuba and demonstrate a strong correlation between local precipitation and multidecadal variability of Atlantic SSTs during the last 1300 years. The authors show that there is no clear drying response in Cuba to either the grand solar minima or volcanic activity and that their record is most likely explained by long-term changes in the AMO.

In Barbuda, the lake sediment record from Freshwater Pond exhibits a more complex response to climate forcing during the LIA. Burn et al. (2016) present an ~400-year reconstruction of effective precipitation for Barbuda since the mid-16th century based on biostratigraphic and stable oxygen isotope analyses of fossil ostracods and gastropods recovered from lake sediment cores (Figure 1.5). Episodic fluctuations in shell accumulation in the sediment record represent changes in the balance between precipitation and evaporation during the LIA and industrial (1850 CE–present) periods. To examine the influence of these modes of climate variability on the long-term effective rainfall in Barbuda, Burn et al. (2016) compared the abundance of the freshwater gastropod *Pyrgophorus parvulus* and an oxygen isotope record from ostracod calcite from Freshwater Pond with high-resolution, tree-ring based reconstructions of the AMO (Gray et al. 2004) and ENSO (Li et al. 2013). The strong correspondence between the abundance of *P. parvulus* and the AMO reconstruction between about 1550 and 1850 CE suggests that extended episodes of aridity were best explained by suppressed SSTs in the tropical Atlantic. Similarly, periods of increased effective moisture were mostly associated with warmer Atlantic SSTs. The covariance between these two proxy records suggests that during the LIA, SST variability in the Atlantic had a greater influence on rainfall patterns in Barbuda than natural radiative forcing, for which no evidence was found. Given Barbuda's proximity to the center of AMO activity in the Tropical Atlantic, the close correspondence between Atlantic SST and rainfall patterns seems intuitive.

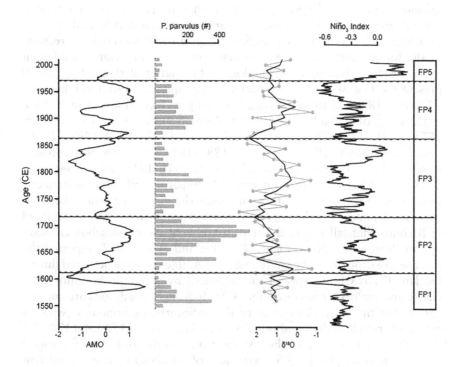

Figure 1.5 400-year reconstruction of effective precipitation from lake sediment cores.

Further comparison of the Freshwater Pond sequence with the ENSO reconstruction of Li et al. (2013; Figure 1.5) suggests that ENSO has also had a strong influence on rainfall patterns in Barbuda during the LIA, particularly since 1700 CE. Episodes of stronger El Niño-like activity occurred during the periods 1720–1775, 1820–1850, and 1975–2010 CE and were associated with drier conditions as represented by the low abundance of *P. parvulus* and more positive $\delta^{18}O$ values. In contrast, inferred wetter periods between 1775 and 1800 CE and during the late 19th century correspond with La Niña-like conditions in the tropical Pacific. Given that contemporary rainfall patterns in the region are strongly controlled by the interplay between ENSO and the AMO, the changing rainfall patterns represented in the Barbuda record also appear to reflect the relative influences of these regional-scale climatic phenomena.

1.3.3.2 Documentary sources from Antigua

Documentary sources of climate variability in the Caribbean include missionary, plantation, and governmental papers and provide climate information at high temporal resolutions that are poorly resolved in most natural environmental archives. Comparison between documental and natural lake sediment archives is difficult to give the chronological precision of the former and the decadal- to centennial-scale uncertainties inherent in most radiocarbon-based chronologies of the latter. Berland et al. (2013) present the first archival investigation of precipitation variability between 1770 and 1890 CE from the island of Antigua, located less than 100 km to the south of Barbuda. They identified significant dry phases in the years 1775–1780, 1788–1791, 1820–1822, 1834–1837, 1844–1845, 1859–1860, 1862–1864, 1870–1874, and 1881–1882 CE, while wet episodes were 1771–1774, 1833–1834, 1837–1838, 1841–1844, 1845–1846, and 1878–1881 CE. Despite the difficulties of comparing archival records of different temporal resolutions, there are some significant similarities between these results and the sediment-based reconstruction from Freshwater Pond in Barbuda that fall within the error margins of the radiocarbon chronology of Burn et al. (2016; Figure 1.5). Berland et al. (2013) reported the most severe drought of the study period in Antigua to have occurred during the late 1770s CE, an episode characterized by severe precipitation deficiency and multi-year harvest failures. Such drought conditions are consistent, within the margins of error of the radiocarbon chronology, with an extended episode of dry conditions in Barbuda between ~1720 and 1780 CE (Unit FP3) characterized by the low abundance of the freshwater gastropod *P. parvulus*, more positive $\delta^{18}O$ excursions of ostracod calcite and persistent El Niño-like conditions in the equatorial Pacific (Burn et al. 2016). In contrast, Berland et al. (2013) report a sustained increase in rainfall from the early 1870s to the mid-1880s CE, which is consistent not only with local meteorological records but also with the rapid transition from a period of extreme aridity recorded at Freshwater Pond at ~1865 CE, similarly characterized by a low abundance of gastropods and positive $\delta^{18}O$ values, to wetter conditions in the 1880s. Such comparisons lend credibility to both reconstructions. Moreover, the sediment-based reconstruction of effective moisture provides the longer-term paleohydrological context within which the more detailed documentary reconstruction may be understood, whereas the documentary source provides specific detail that is oftentimes poorly resolved in the sediment record.

1.4 Future climate projections

Projected climate change in the 21st century is of significant importance for the SIDS of the Caribbean, which is extremely vulnerable to multiple climate stressors including sea level rise (SLR), the passage of tropical cyclones, increasing air and SSTs, and extended episodes of drought (Nurse et al. 2014). Here, we review the global climate projections from the Fifth Assessment Report (AR5) of the Intergovernmental Panel on Climate

Change (IPCC) and discuss how these projected climatic changes are manifested at the regional (Caribbean-wide) and local (Antigua and Barbuda) scales.

1.4.1 Global climate projections

The AR5 of the IPCC identifies four Representative Concentration Pathways (RCPs) for different projections of future greenhouse gas (GHG) concentrations. RCP 2.6 assumes that global annual GHG concentrations peak between 2010 and 2020 CE at 3 Wm^{-2}, declining substantially thereafter; those in RCP 4.5 peak and stabilize at 4.5 Wm^{-2} around 2040 CE; in RCP 6.0, GHGs peak and stabilize around 2080 CE; and lastly, in RCP 8.5, concentrations reach 8.5 Wm^{-2} by 2100 CE and continue to rise thereafter (Cubasch et al. 2013). In general, RCP 4.5 is used by most climate practitioners and represents an intermediate total radiative forcing pathway coupled with a positive global GHG forcing of 4.5 Wm^{-2} by 2100. It is a mitigation-based, and arguably more realistic, pathway that assumes positive human intervention, such as global uptake of renewable energy sources to avoid a runaway warming climate, of which the latter is represented by the fossil-fuel intensive RCP 8.5 scenario with a positive GHG forcing of 8.5 Wm^{-2} by 2100 (Taylor et al. 2018).

In terms of temperature, the average mean global temperature is projected to exceed 1.5°C for all RCP scenarios relative to 1850–1900 CE, 2°C for RCP 6.0 and RCP 8.5, and more likely than not 2°C for RCP 2.6. As a result, global sea levels have been rising at a rate of 3.2 mm per year since 1993 (NOAA 2018), although there are local and regional differences due to the consequences of the vertical land movement and the eustatic sea level due to thermal expansion and melting land glaciers. The global mean sea level will continue to rise in the 21st century at a faster rate. The amount of rising varies from 0.26–0.55 m for RCP 2.6 to 0.45–0.82 m for RCP 8.5 and between 0.4 and 0.7 m for the RCP 4.5 scenario (Christensen et al. 2013).

1.4.2 Future climate projections for the Caribbean

Taylor et al. (2018) present projections of temperature and precipitation for the Caribbean region based on three global warming scenarios: projections for a world whose pre-industrial mean temperature (1861–1900) has been exceeded by global mean annual air temperature anomalies of 1.5°C, 2°C, and 2.5°C (designated $\Delta T_g 1.5$, $\Delta T_g 2$, $\Delta T_g 2.5$; Taylor et al. 2018). They simulate near-surface temperature and precipitation values using ten General Circulation Models (GCMs) from the Coupled Model Intercomparison Project phase 5 (CMIP5) for RCP 4.5. Each of the ten selected GCMs was able to capture the characteristic bimodal rainfall pattern of the Caribbean (Figure 1.3) and is considered to be a good representation of the regional climatology. Here, we summarize the future temperature and precipitation states for the Caribbean using RCP 4.5 and for the global warming scenarios $\Delta T_g 1.5$, $\Delta T_g 2.0$, and $\Delta T_g 2.5$.

When compared with the pre-industrial baseline, temperature projections for the Caribbean show a significant increase in mean annual temperature anomalies of 1.2 ± 0.2°C ($\Delta T_g 1.5$) by 2028, 1.6 ± 0.3°C ($\Delta T_g 2.0$) by 2046, and 2.0 ± 0.3°C ($\Delta T_g 2.5$) by 2070 (Taylor et al. 2018). These estimates are consistent with previous regional model projections, which show temperature increases spanning 1.0–3.5°C by the end of the 21st century (Campbell et al. 2011; McSweeney et al. 2012; Karmalkar et al. 2013). Regional variability is expressed by enhanced (reduced) warming on larger (smaller) land masses such as the Greater (Lesser) Antilles. The associated sea level rise in the Caribbean region is estimated to be close to the global mean at 0.5–0.6 m by 2100 for RCP 4.5 (Nurse et al. 2014).

The projected precipitation regime for the region is characterized by significant spatial and temporal variability, which resembles the complexity of past and present climate systems of the Caribbean described above. Under the $\Delta T_g 1.5$ scenario, much of the region is expected to show an increase in precipitation, with a maximum increase in the southwestern Caribbean associated with enhanced moisture transport within the CLLJ (Taylor et al. 2018). In contrast, the north- and southeast Caribbean islands, including the islands ranging from Puerto Rico to Antigua and Barbuda as well as the Netherlands Antilles, are expected to experience a slightly drier climate. Under the $\Delta T_g 2.0$ and $\Delta T_g 2.5$ scenarios, the entire region, with the exception of the Bahamas, exhibits a drying trend of between 5% and 20%, with spatial variability still being significant. Projected changes in rainfall intensity correspond to those of precipitation anomalies where the number of days of the year when precipitation exceeds 10 mm increase for $\Delta T_g 1.5$ and then decrease progressively for $\Delta T_g 2.0$ and $\Delta T_g 2.5$ (Taylor et al. 2018). Given the lack of a concerted effort toward a low-carbon global economy, a pragmatic approach would adopt the $\Delta T_g 2.0$ and $\Delta T_g 2.5$ scenarios as the most realistic global warming targets. Consequently, under the current GHG trajectory, it is likely that the Caribbean will experience a progressive drying trend by 2100 that is characterized by a more intense CLLJ and fewer intense rainfall days.

The impact of 21st-century climate change on Atlantic tropical cyclones is poorly understood. The IPCC AR5 (2013) report, informed by the work of Knutson et al. (2010), concluded that there is low confidence in region-specific projections of tropical cyclone activity and that it remains uncertain whether recent changes in Atlantic tropical cyclone activity lie outside the range of natural variability (Christensen et al. 2013). Such uncertainty makes it difficult for SIDS to plan for the significant socio-economic and cultural consequences of 21st-century global warming. Nevertheless, there is a broad scientific consensus that there will be a substantial decrease (ca. 25%) in the frequency of tropical cyclones, which is consistent with 21st-century projections of reduced precipitation, a stronger CLLJ, and enhanced vertical wind shear in the tropical Atlantic. Given the anthropogenic warming of Atlantic SSTs, however, it is considered likely that storms will become more intense and exhibit higher rainfall rates than contemporary tropical cyclones (Bender et al. 2010). Thus, the likely response of Atlantic tropical cyclones to anthropogenic climate change is a transition to fewer than average tropical cyclones with greater intensity. Arguably, SIDS of the Caribbean should therefore assign greater importance to the expected increased vulnerability of their coastlines to greater storm surge flooding associated with future sea level rise and coastal development.

1.4.3 Impact of projected 21st-century climate change on ENSO and the AMO

Given the strong influence of ENSO and the AMO on climate phenomena in the Caribbean, the future dynamics of coupled ocean–atmosphere variability in the Pacific and Atlantic Oceans needs to be considered when projecting 21st-century climate. However, the extent to which these modes of climate variability respond to anthropogenic climate change is poorly understood. Contemporary SST anomalies in the tropical Atlantic have shown a consistent negative trend since 2013, prompting Klotzbach et al. (2015) to suggest a transition may be underway from the positive AMO phase of enhanced rainfall and tropical cyclone activity that characterized the period 1995–2012 to a negative phase that would portend less rainfall and tropical cyclone activity for at least half an AMO cycle (Goldenberg et al. 2001). Such conditions superimposed on the projected progressive drying trend linked to 21st-century anthropogenic warming would act to further suppress rainfall and tropical cyclone activity.

The response of ENSO to 21st-century climate change is equally poorly understood; however, recent studies have provided further insights into the link between the changing mean state of the eastern equatorial Pacific and ENSO dynamics. In a comprehensive review of empirical and climate model data, Cai et al. (2015) showed that the accelerated warming of eastern equatorial Pacific SSTs associated with a reduced Walker circulation is expected to increase extreme rainfall and the equatorward migration of the Pacific ITCZ, both of which are features of an extreme El Niño event. The authors argue further that the frequency of extreme La Niña events is also expected to increase and that ENSO-related extreme weather events are likely to occur more frequently under 21st-century global warming. In the Caribbean, this would manifest itself as a greater magnitude of interannual variability between episodes of extreme rainfall and tropical cyclone activity and those of drought. Arguably, the transition from the extreme Caribbean-wide droughts of the 2015–2016 El Niño to the very active and record-breaking Atlantic hurricane seasons of 2017 and 2020 (both La Niña years) is consistent with this hypothesized pattern of extreme ENSO-related climate phenomena.

1.4.4 Future climate change in Barbuda

The impact of SLR is likely not only to exacerbate coastal flooding and the concomitant erosion of mangrove and coral reef ecosystems but also to enhance the devastation from storm surges associated with the passage of tropical cyclones and tsunamis, as well as promote saline intrusion and the associated degradation of potable groundwater resources, a particular problem for islands composed of porous limestone geology. The impact of the rising sea levels on Barbuda is already unmistakable, characterized by enhanced levels of saline intrusion (Boger et al. 2014), greater susceptibility to hurricane-induced marine washover events, the consequent harmful effects of elevated soil salinity (Boger et al. 2016), and the increasingly rapid erosion of terrestrial and marine ecosystems as well as archaeological sites located along the island's coastlines (Perdikaris et al. 2018). A recent report from the Permanent Service for Mean Sea Level (PSMSL) estimated a negative vertical land movement of –0.070 mm per year for Barbuda, which approximates ~0.5 m of SLR by the year 2100, a more conservative estimate of the extent of the impact from SLR compared with the 0.7 m SLR calculated by the IPCC AR5 report (Christensen et al. 2013). Under a 0.5 m SLR scenario, Barbuda would lose ~30% of its total land area, as shown by a recent Digital Elevation Model (DEM) of SLR (Figure 1.6). In particular, the eastern and northern coastal low-lying areas are particularly susceptible to the impacts of SLR, including areas in close proximity to Codrington, the island's only village.

Available freshwater has always been a significant challenge in Barbuda due to its low topography, limited rainfall, porous limestone geology, and limited capacity to store water. Given the projections of regional drying during the 21st century, Barbuda is also likely to experience drier conditions in the immediate future under $\Delta T_g 1.5$, $\Delta T_g 2.0$, and $\Delta T_g 2.5$ global warming scenarios. Superimposed on this trend, the island may experience extended periods of drought associated with the negative phases of the AMO and pronounced periods of extreme rainfall and drought associated with the increased amplitude of variability of the ENSO cycle (Cai et al. 2015; Klotzbach 2011). In 2018, Barbuda experienced extreme drought leading the Caribbean Climate Outlook Forum (CariCOF) to put a drought watch in place for Antigua and Barbuda for long-term drought conditions through the winter of 2018/2019. Subsequent El Niño conditions extended from September 2018 through May 2019, causing widespread drought. Given that Elder et al. (2014) found a positive correlation of drought conditions between the AMO and the Standardized Precipitation Evapotranspiration Index (SPEI; negative values of SPEI indicate drought), it appears that combined AMO and El Niño-like conditions contributed to the severe drought that Barbuda faced in 2018/2019. Taken together, the devastating impacts of Hurricane Irma in 2017 (discussed below) combined with an extended drought in 2018/2019 had a significant impact on both terrestrial and marine ecosystems of Barbuda as well as its intangible and tangible heritage

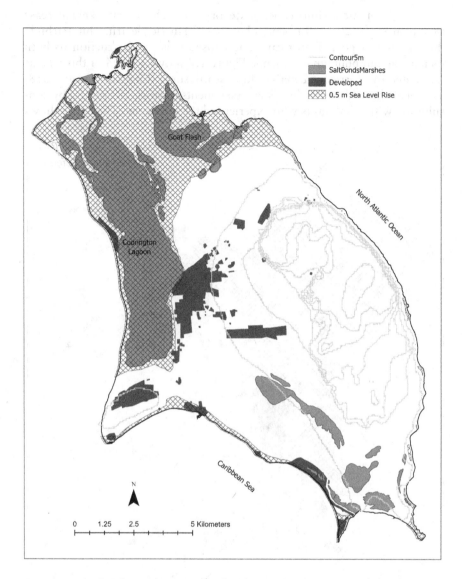

Figure 1.6 Sea level rise of 0.5 m. Inundated areas are shown in cross-hatched.

(Boger et al. 2019), which could mark a profound change in social, cultural, and ecological dynamics.

In addition to the direct loss of land due to SLR, large areas of Barbuda are also affected by the storm surges associated with the passage of tropical cyclones. While the frequency and intensity of future storms are poorly resolved in global and regional climate models, it is generally thought that the number of intense hurricanes (Category 3 and above) will likely increase (Knutson et al. 2010). In September 2017, Hurricane Irma hit Barbuda directly as Category 5 hurricane and caused massive destruction to both its natural and built environments. Figure 1.7 shows a DEM of those areas inundated by Hurricane Irma's 2.43 m storm surge (Cangialosi et al. 2018) as well as those identified by visual assessments of storm debris and by consultation with inhabitants who experienced the hurricane. The figure shows

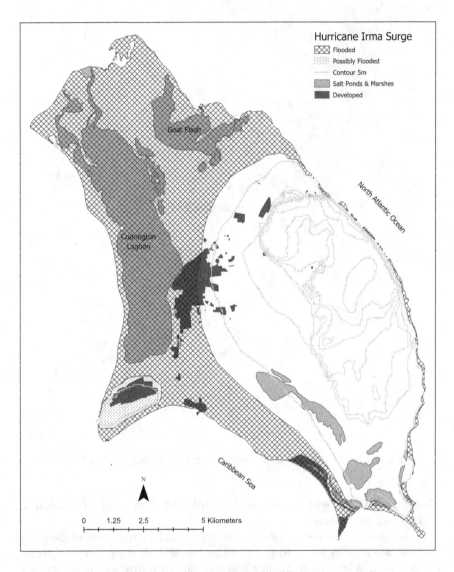

Figure 1.7 Storm surge estimates for Hurricane Irma.

that a large proportion of the total population of Barbuda who live in the village of Codrington was affected by the surge. Indeed, the combined damage of storm surge, rain, and wind resulted in 90–95% or more of the built environment being severely impacted by the hurricane (Cangialosi et al. 2018). Considering the projected 21st-century SLR for Barbuda, similar magnitude storm surges in the future will likely be more devastating as they penetrate further inland. Furthermore, recent research by Rivera-Collazo (2020) highlights the disparity between predictive modeling of climatic impacts and the actual damage to local cultural and heritage resources. These are important considerations as Barbuda recovers from Hurricane Irma and explores ways to increase resiliency to projected 21st-century hurricane activity.

The increased salinization of surface-fed groundwaters and soils owing to the washover of seawater during a hurricane is a further challenge to future agriculture in Barbuda and is the subject of ongoing scientific research on the island (Boger et al. 2016). Indeed, initial soil salinity measurements taken in January 2018 show high salinity values >3 mS/cm (R. Boger; unpublished data) in the surge area. When combined with the likely drop in effective precipitation associated with projected 21st-century drying, the increased salinity of soil and groundwater will likely result in lower levels of potable fresh water in the coming decades than was previously estimated. Large stretches of agricultural land were affected by the salinization associated with the storm surge from Hurricane Irma. As a result, inhabitants may increasingly need to grow crops that are better adapted to saline soils as well as adopt new agricultural techniques designed for growing in saline soil and brackish groundwater (Boger et al. 2016).

1.5 Conclusions

The aim of this chapter was to examine the nature of Barbuda's climate within the broader context of the climate dynamics of the Caribbean region and beyond. Guided by the principles of uniformitarianism, we emphasize that a comprehensive understanding of contemporary climate dynamics of the Caribbean is essential to inform us not only about the history of the climate of the region but also about the possible trajectories of future climate scenarios. From an archaeologist's perspective, such knowledge facilitates direct comparisons between the histories of human settlement and climate and helps us decipher the extent to which human activities and migration were determined climatically. However, the evidence that underpins our knowledge of contemporary and past climate variability within the region is sporadic and further collection and analyses of climate and paleoclimate data are required before we can truly understand the spatial and temporal complexity of Caribbean climate dynamics and its potential relationship with human activity in the past, present, and future.

References

Angeles, M.E., González, J.E., Ramírez-Beltrán, N.D., Tepley, C.A. and Comarazamy, D.E., 2010. Origins of the Caribbean rainfall bimodal behavior. *Journal of Geophysical Research: Atmospheres*, 115(D11). https://agupubs. onlinelibrary.wiley.com/doi/10.1029/2009JD012990.

Bain, A., Faucher, A.M., Kennedy, L.M., LeBlanc, A.R., Burn, M.J., Boger, R. and Perdikaris, S., 2018. Landscape transformation during ceramic age and colonial occupations of Barbuda, West Indies. *Environmental Archaeology*, 23(1), pp.36–46.

Bender, M.A., Knutson, T.R., Tuleya, R.E., Sirutis, J.J., Vecchi, G.A., Garner, S.T. and Held, I.M., 2010. Modeled impact of anthropogenic warming on the frequency of intense Atlantic hurricanes. *Science*, 327(5964), pp.454–458.

Berland, A.J. and Endfield, G., 2018. Drought and disaster in a revolutionary age: Colonial Antigua during the American independence war. *Environment and History*, 24(2), pp.209–235.

Berland, A.J., Metcalfe, S.E. and Endfield, G.H., 2013. Documentary-derived chronologies of rainfall variability in Antigua, Lesser Antilles, 1770–1890. *Climate of the Past*, 9(3), pp.1331–1343.

Bindoff, N.L, Stott, P.A., AchutaRao, K.M., Allen, M.R., Gillett, N., Gutzler, D., Hansingo, K., Hegerl, G., Hu, Y., Jain, S., Mokhov, I.I., Overland, J., Perlwitz, J., Sebbari, R. and Zhang, X., 2013. Climate Phenomena and their Relevance for Future Regional Climate Change. In *Climate Change 2013 – The Physical Science Basis*, Intergovernmental Panel on Climate Change, pp.1217–1308. Cambridge University Press, Cambridge.

Black, D.E., Abahazi, M.A., Thunell, R.C., Kaplan, A., Tappa, E.J. and Peterson, L.C., 2007. An 8-century tropical Atlantic SST record from the Cariaco Basin: Baseline variability, twentieth-century warming, and Atlantic hurricane frequency. *Paleoceanography*, 22(4). https://agupubs.onlinelibrary.wiley.com/doi/full/10.1029/2007PA001427.

Boger, R., Perdikaris, S., Potter, A.E. and Adams, J., 2016. Sustainable resilience in Barbuda: Learning from the past and developing strategies for the future. *International Journal of Environmental Sustainability*, 12(4), p.1.

Boger, R., Perdikaris, S., Potter, A.E., Mussington, J., Murphy, R., Thomas, L., Gore, C. and Finch, D., 2014. Water resources and the historic wells of Barbuda: Tradition, heritage and hope for a sustainable future. *Island Studies Journal*, 9(2), pp. 327–342.

Boger, R., Perdikaris, S. and Rivera-Collazo, I., 2019. Cultural heritage and local ecological knowledge under threat: Two Caribbean examples from Barbuda and Puerto Rico. *Journal of Anthropology and Archaeology*, 7(2), pp.1–14.

Brasier, M. and Donahue, J., 1985. Barbuda—an emerging reef and lagoon complex on the edge of the Lesser Antilles island are. *Journal of the Geological Society*, 142(6), pp.1101–1117.

Broecker, W.S., 2000. Was a change in thermohaline circulation responsible for the Little Ice Age?. *Proceedings of the National Academy of Sciences*, 97(4), pp.1339–1342.

Burn, M.J., Holmes, J., Kennedy, L.M., Bain, A., Marshall, J.D. and Perdikaris, S., 2016. A sediment-based reconstruction of Caribbean effective precipitation during the 'Little Ice Age'from Freshwater Pond, Barbuda. *The Holocene*, 26(8), pp.1237–1247.

Burn, M.J. and Palmer, S.E., 2014. Solar forcing of Caribbean drought events during the last millennium. *Journal of Quaternary Science*, 29(8), pp.827–836.

Cai, W., Santoso, A., Wang, G., Yeh, S.W., An, S.I., Cobb, K.M., Collins, M., Guilyardi, E., Jin, F.F., Kug, J.S. and Lengaigne, M., 2015. ENSO and greenhouse warming. *Nature Climate Change*, 5(9), pp.849–859.

Campbell, J.D., Taylor, M.A., Stephenson, T.S., Watson, R.A. and Whyte, F.S., 2011. Future climate of the Caribbean from a regional climate model. *International Journal of Climatology*, 31(12), pp.1866–1878.

Cangialosi, John P., Latto, Andrew S. and Berg, Robbie., 2018. *National Hurricane Center Tropical Cyclone Report: Hurricane Irma (AL12017), 30 August–12 September 2017*. National Hurricane Center.

Chen, A. Anthony and Taylor, Michael A., 2002. Investigating the link between early season Caribbean rainfall and the El Niño+1 year. *International Journal of Climatology*, 22(1), pp.87–106.

Chenoweth, M., 2007. Objective classification of historical tropical cyclone intensity. *Journal of Geophysical Research: Atmospheres*, 112(D5). https://agupubs.onlinelibrary.wiley.com/doi/full/10.1029/2006JD007211.

Chenoweth, M. and Divine, D., 2008. A document-based 318-year record of tropical cyclones in the Lesser Antilles, 1690–2007. *Geochemistry, Geophysics, Geosystems*, 9(8). https://agupubs.onlinelibrary.wiley.com/doi/full/10.1029/200 8GC002066.

Chenoweth, M. and Divine, D., 2014. Eastern Atlantic tropical cyclone frequency from 1851–1898 is comparable to satellite era frequency. *Environmental Research Letters*, 9(11), p.114023.

Christensen, J.H., Kumar, K., Aldrian, E., An, S.-I., Cavalcanti, I.F.A., de Castro, M., Dong, W., Goswami, P., Hall, A., Kanyanga, J.K., Kitoh, A., Kossin, J., Lau, N.-C., Renwick, J., Stephenson, D.B., Xie, S.-P. and Zhou, T., 2013. Climate phenomena and their relevance for future regional climate change. In *Climate Change 2013: The Physical Science Basis. Contribution of Working Group I to the Fifth Assessment Report of the Intergovernmental Panel on Climate Change*, edited by V. Bex and P.M. Midgley Stocker, T.F., D. Qin, G.-K. Plattner, M. Tignor, S.K. Allen, J. Boschung, A. Nauels, and Y. Xia. Cambridge University Press, Cambridge, United Kingdom and New York.

Cook, K.H. and Vizy, E.K., 2010. Hydrodynamics of the Caribbean low-level jet and its relationship to precipitation. *Journal of Climate*, 23(6), pp.1477–1494.

Cubasch, Ulrich, Wuebbles, Donald, Chen, Deliang, Facchini, Maria Cristina, Frame, David, Mahowald, Natalie and Winther, Jan-Gunnar, 2013. Introduction in climate change 2013. In *Intergovernmental Panel on Climate Change 2013: The Physical Science Basis. Contribution of Working Group I to the Fifth Assessment Report of the Intergovernmental Panel on Climate Change*, edited by Stocker, T.F., D. Qin, G.-K. Plattner, M. Tignor, S.K. Allen, J. Boschung, A. Nauels, Y. Xia, V. Bex and P.M. Midgley. Cambridge University Press, Cambridge, United Kingdom and New York, NY, USA pp.119–158.

Curtis, J.H., Brenner, M. and Hodell, D.A., 2001. Climate change in the Circum-Caribbean (Late Pleistocene to Present) and implications for regional biogeography. *Biogeography of the West Indies: Patterns and Perspectives*, 2, pp.35–54.

Curtis, S., 2013. Daily precipitation distributions over the intra-Americas sea and their interannual variability. *Atmósfera*, 26(2), pp.243–259.

Curtis, S. and Gamble, D.W., 2008. Regional variations of the Caribbean mid-summer drought. *Theoretical and Applied Climatology*, 94(1), pp.25–34.

Elder, R.C., Balling, R.C., Cerveny, R.S. and Krahenbuhl, D., 2014. Regional variability in drought as a function of the Atlantic Multidecadal Oscillation. *Caribbean Journal of Science*, 48(1), pp.31–43.

Enfield, D.B., Mestas-Nuñez, A.M. and Trimble, P.J., 2001. The Atlantic multidecadal oscillation and its relation to rainfall and river flows in the continental US. *Geophysical Research Letters*, 28, pp.2077–2080.

Fairbanks, R.G., 1989. A 17,000-year glacio-eustatic sea level record: Influence of glacial melting rates on the Younger Dryas event and deep-ocean circulation. *Nature*, 342(6250), pp.637–642.

Fensterer, C., Scholz, D., Hoffmann, D., Spötl, C., Pajón, J.M. and Mangini, A., 2012. Cuban stalagmite suggests relationship between Caribbean precipitation and the Atlantic Multidecadal Oscillation during the past 1.3 ka. *The Holocene*, 22(12), pp.1405–1412.

Fensterer, C., Scholz, D., Hoffmann, D.L., Spötl, C., Schröder-Ritzrau, A., Horn, C., Pajon, J.M. and Mangini, A., 2013. Millennial-scale climate variability during the last 12.5 ka recorded in a Caribbean speleothem. *Earth and Planetary Science Letters*, 361, pp.143–151.

Folland, C.K., Palmer, T.N. and Parker, D.E., 1986. Sahel rainfall and worldwide sea temperatures, 1901–85. *Nature*, 320(6063), pp.602–607.

Fritz, S.C., Björck, S., Rigsby, C.A., Baker, P.A., Calder-Church, A. and Conley, D.J., 2011. Caribbean hydrological variability during the Holocene as reconstructed from crater lakes on the island of Grenada. *Journal of Quaternary Science*, 26(8), pp.829–838.

Gamble, D.W. and Curtis, S., 2008. Caribbean precipitation: Review, model and prospect. *Progress in Physical Geography*, 32(3), pp.265–276.

Gamble, D.W., Parnell, D.B. and Curtis, S., 2008. Spatial variability of the Caribbean mid-summer drought and relation to north Atlantic high circulation. *International Journal of Climatology: A Journal of the Royal Meteorological Society*, 28(3), pp.343–350.

García-Herrera, R., Gimeno, L., Ribera, P. and Hernández, E., 2005. New records of Atlantic hurricanes from Spanish documentary sources. *Journal of Geophysical Research: Atmospheres*, 110(D3). https://agupubs.onlinelibrary.wiley.com/doi/10.1029/2004JD005272.

Goldenberg, S.B., Landsea, C.W., Mestas-Nuñez, A.M. and Gray, W.M., 2001. The recent increase in Atlantic hurricane activity: Causes and implications. *Science*, 293(5529), pp.474–479.

Gouirand, I., Jury, M.R. and Sing, B., 2012. An analysis of low-and high-frequency summer climate variability around the Caribbean Antilles. *Journal of Climate*, 25(11), pp.3942–3952.

Gray, S.T., Graumlich, L.J., Betancourt, J.L. and Pederson, G.T., 2004. A tree-ring based reconstruction of the Atlantic Multidecadal Oscillation since 1567 AD. *Geophysical Research Letters*, 31(12). https://agupubs.onlinelibrary.wiley.com/doi/10.1029/2004GL019932.

Gray, W.M., 1968. Global view of the origin of tropical disturbances and storms. *Monthly Weather Review*, 96(10), pp.669–700.

Haase-Schramm, A., Böhm, F., Eisenhauer, A., Dullo, W.C., Joachimski, M.M., Hansen, B. and Reitner, J., 2003. Sr/Ca ratios and oxygen isotopes from sclerosponges: Temperature history of the Caribbean mixed layer and thermocline during the Little Ice Age. *Paleoceanography, 18*(3), p.1073.

Haug, Gerald H., Hughen, Konrad A., Sigman, Daniel M., Peterson, Larry C. and Röhl, U., 2001. Southward migration of the intertropical convergence zone through the Holocene. *Science, 293*(5533), pp.1304–1308.

Hetzinger, S., Pfeiffer, M., Dullo, W.C., Keenlyside, N., Latif, M. and Zinke, J., 2008. Caribbean coral tracks Atlantic Multidecadal Oscillation and past hurricane activity. *Geology, 36*(1), pp.11–14.

Higuera-Gundy, A., Brenner, M., Hodell, D.A., Curtis, J.H., Leyden, B.W. and Binford, M.W., 1999. A 10,300 14C yr record of climate and vegetation change from Haiti. *Quaternary Research, 52*(2), pp.159–170.

Hodell, D.A., Brenner, M. and Curtis, J.H., 2005. Terminal Classic drought in the northern Maya lowlands inferred from multiple sediment cores in Lake Chichancanab (Mexico). *Quaternary Science Reviews, 24*(12–13), pp.1413–1427.

Hodell, D.A., Curtis, J.H., Jones, G.A., Higuera-Gundy, A., Brenner, M., Binford, M.W. and Dorsey, K.T., 1991. Reconstruction of Caribbean climate change over the past 10,500 years. *Nature, 352*(6338), pp.790–793.

Holmes, J.A., 1997. Recent non-marine Ostracoda from Jamaica, West Indies. *Journal of Micropalaeontology, 16*(2), pp.137–143.

Holmes, J.A., Street-Peffott, F.A., Ivanovich, M. and Peffott, R.A., 1995. A late quaternary palaeolimnological record from Jamaica based on trace-element chemistry of ostracod shells. *Chemical Geology, 124*(1–2), pp.143–160.

Holzhauser, H., Magny, M. and Zumbuühl, H.J., 2005. Glacier and lake-level variations in west-central Europe over the last 3500 years. *The Holocene, 15*(6), pp.789–801.

Jackson, I., 2001. *Drought Hazard Assessment and Mapping for Antigua and Barbuda: Post-Georges Disaster Mitigation Project in Antigua & Barbuda and St. Kitts & Nevis, April 2001*. Organization of American States, Unit for Sustainable Development and Environment.

Jones, P.D., Briffa, K.R., Osborn, T.J., Lough, J.M., van Ommen, T.D., Vinther, B.M., Luterbacher, J., Wahl, E.R., Zwiers, F.W., Mann, M.E. and Schmidt, G.A., 2009. High-resolution palaeoclimatology of the last millennium: A review of current status and future prospects. The Holocene, 19(1), pp.3–49.

Jury, M., Malmgren, B.A. and Winter, A., 2007. Subregional precipitation climate of the Caribbean and relationships with ENSO and NAO. *Journal of Geophysical Research: Atmospheres, 112*(D16). https://agupubs.onlinelibrary.wiley.com/doi/full/10.1029/2006JD007541.

Karmalkar, A.V., Taylor, M.A., Campbell, J., Stephenson, T., New, M., Centella, A., Benzanilla, A. and Charlery, J., 2013. A review of observed and projected changes in climate for the islands in the Caribbean. *Atmósfera, 26*(2), pp.283–309.

Keegan, W.F., Hofman, C.L. and Rodríguez Ramos, R., 2013. Introduction. In *The Oxford Handbook of Caribbean Archaeology*, edited by W.F. Keegan, C.L. Hofman, and R. Rodríguez Ramos, pp.1–18. Oxford University Press, New York.

Keegan, W.F. and C.L. Hofman., 2017. *The Caribbean Before Columbus*. New York: Oxford University Press.

Keigwin, L.D., 1996. The little ice age and medieval warm period in the Sargasso Sea. *Science, 274*(5292), pp.1504–1508.

Kennett, D.J., Breitenbach, S.F., Aquino, V.V., Asmerom, Y., Awe, J., Baldini, J.U., Bartlein, P., Culleton, B.J., Ebert, C., Jazwa, C. and Macri, M.J., 2012. Development and disintegration of Maya political systems in response to climate change. *Science*, *338*(6108), pp.788–791.

Kilbourne, K.H., Quinn, T.M., Webb, R., Guilderson, T., Nyberg, J. and Winter, A., 2008. Paleoclimate proxy perspective on Caribbean climate since the year 1751: Evidence of cooler temperatures and multidecadal variability. *Paleoceanography*, *23*(3). https://agupubs.onlinelibrary.wiley.com/doi/full/10.10 29/2008PA001598.

Klotzbach, P., Gray, W. and Fogarty, C., 2015. Active Atlantic hurricane era at its end?. *Nature Geoscience*, *8*(10), pp.737–738.

Klotzbach, P.J., 2011. The influence of El Niño–Southern Oscillation and the Atlantic multidecadal oscillation on Caribbean tropical cyclone activity. *Journal of Climate*, *24*(3), pp.721–731.

Knight, J.R., Folland, C.K. and Scaife, A.A., 2006. Climate impacts of the Atlantic multidecadal oscillation. *Geophysical Research Letters*, *33*(17). https://agupubs. onlinelibrary.wiley.com/doi/10.1029/2006GL026242.

Knutson, T.R., McBride, J.L., Chan, J., Emanuel, K., Holland, G., Landsea, C., Held, I., Kossin, J.P., Srivastava, A.K. and Sugi, M., 2010. Tropical cyclones and climate change. *Nature Geoscience*, *3*(3), pp.157–163.

Kossin, J.P. and Vimont, D.J., 2007. A more general framework for understanding Atlantic hurricane variability and trends. *Bulletin of the American Meteorological Society*, *88*(11), pp.1767–1782.

Landsea, C., 1993. A climatology of intense (or major) Atlantic hurricanes. *Monthly Weather Reviews*, *121*, pp.1703–1713.

Lane, C.S., Horn, S.P., Orvis, K.H. and Thomason, J.M., 2011. Oxygen isotope evidence of Little Ice Age aridity on the Caribbean slope of the Cordillera Central, Dominican Republic. *Quaternary Research*, *75*(3), pp.461–470.

LeFebvre, M.J., Giovas, C.M. and Laffoon, J.E., 2019. Advancing the study of Amerindian ecodynamics in the Caribbean: Current perspectives. *Environmental Archaeology*, *24*(2), pp.104–114.

Li, J., Xie, S.P., Cook, E.R., Morales, M.S., Christie, D.A., Johnson, N.C., Chen, F., D'Arrigo, R., Fowler, A.M., Gou, X. and Fang, K., 2013. El Niño modulations over the past seven centuries. *Nature Climate Change*, *3*(9), pp.822–826.

Magaña, V., Amador, J.A. and Medina, S., 1999. The midsummer drought over Mexico and Central America. *Journal of Climate*, *12*(6), pp.1577–1588.

Mann, M.E., Zhang, Z., Rutherford, S., Bradley, R.S., Hughes, M.K., Shindell, D., Ammann, C., Faluvegi, G. and Ni, F., 2009. Global signatures and dynamical origins of the Little Ice Age and Medieval Climate Anomaly. *Science*, *326*(5957), pp.1256–1260.

Marchitto, T.M. and Demenocal, P.B., 2003. Late Holocene variability of upper North Atlantic Deep Water temperature and salinity. *Geochemistry, Geophysics, Geosystems*, *4*(12). https://agupubs.onlinelibrary.wiley.com/doi/full/10.1029/20 03GC000598.

Martin, E.R. and Schumacher, C., 2011. The Caribbean low-level jet and its relationship with precipitation in IPCC AR4 models. *Journal of Climate*, *24*(22), pp.5935–5950.

Martin-Kaye, P.H.A., 1959. *Reports on the Geology of the Leeward and British Virgin Islands*. Voice Publishing Co. Ltd, Port Castries, St Lucia.

McSweeney, C.F., Jones, R.G. and Booth, B.B., 2012. Selecting ensemble members to provide regional climate change information. *Journal of Climate*, 25(20), pp.7100–7121.

Metcalfe, S.E., Jones, M.D., Davies, S.J., Noren, A. and MacKenzie, A., 2010. Climate variability over the last two millennia in the North American Monsoon region, recorded in laminated lake sediments from Laguna de Juanacatlán, Mexico. *The Holocene*, 20(8), pp.1195–1206.

Moron, V., Gouirand, I. and Taylor, M., 2016. Weather types across the Caribbean basin and their relationship with rainfall and sea surface temperature. *Climate Dynamics*, 47(1), pp.601–621.

Napolitano, M.F., DiNapoli, R.J., Stone, J.H., Levin, M.J., Jew, N.P., Lane, B.G., O'Connor, J.T. and Fitzpatrick, S., 2019. Reevaluating human colonization of the Caribbean using chronometric hygiene and Bayesian modeling. *Science Advances*, 5(12), eaar7806 DOI:10.1126/sciadv.aar7806.

Newsom, L. and Wing, E.S., 2004. *Land and Sea: Native American Uses of Biological Resources in the Caribbean*. University of Alabama Press, Tuscaloosa.

NOAA, 2018. https://www.noaa.gov/education/resource-collections/climate/climate-change-impacts.

Nurse, L.A., McLean, R.F., Trinidad, J.A., Briguglio, L.P., Duvat-Magnan, V., Pelesikoti, N., Tompkins, E. and Arthur Webb, A., 2014. Small Islands. *Climate Change 2014: Impacts, Adaptation, and Vulnerability. Part B: Regional Aspects. Contribution of working Group II to the Fifth Assessment Report of the Intergovernmental Panel on Climate Change*, pp.1613–1654.

Nyberg, J., Malmgren, B.A., Kuijpers, A. and Winter, A., 2002. A centennial-scale variability of tropical North Atlantic surface hydrography during the late Holocene. *Palaeogeography, Palaeoclimatology, Palaeoecology*, 183(1–2), pp.25–41.

Perdikaris, S., Bain, A., Boger, R., Grouard, S., Faucher, A.M., Rousseau, V., Persaud, R., Noel, S., Brown, M. and Medina-Triana, J., 2018. Cultural heritage under threat: The effects of Climate Change on the small island of Barbuda, Lesser Antilles. R. In *Public Archaeology and Climate Change*, edited by T. Dawson, C. Nimura, E. López-Romero and M.Y. Daire, pp.138–148. Oxbow Books, Oxford.

Perdikaris, S., Boger, R., and Ibrahimpasic, E., 2021. Seduction, promises and the disneyfication of barbuda post irma. *TRANSLOCAL Contemporary Local and Urban Cultures Journal*. Number 5 (un)inhabited spaces. Ana Salgueiro and Nuno Marques (eds). ISSN 2184-1519 Madeira and Umeå.

Polissar, P.J., Abbott, M.B., Wolfe, A.P., Bezada, M., Rull, V. and Bradley, R.S., 2006. Solar modulation of Little Ice Age climate in the tropical Andes. *Proceedings of the National Academy of Sciences*, 103(24), pp.8937–8942.

Rivera-Collazo, I.C., 2020. Severe weather and the reliability of desk-based vulnerability assessments: the impact of Hurricane Maria to Puerto Rico's coastal archaeology. *The Journal of Island and Coastal Archaeology*, 15(2), pp.244–263.

Rodbell, D.T., Seltzer, G.O., Anderson, D.M., Abbott, M.B., Enfield, D.B. and Newman, J.H., 1999. An~ 15,000-year record of El Niño-driven alluviation in southwestern Ecuador. *Science*, 283(5401), pp.516–520.

Rousseau, V., Bain, A., Chabot, J., Grouard, S. and Perdikaris, S., 2017. The role of Barbuda in the settlement of the Leeward Islands: Lithic and shell analysis along the strombus line shell midden. *Journal of Caribbean Archaeology*, 17(1), pp. 1–25.

Saenger, C., Cohen, A.L., Oppo, D.W., Halley, R.B. and Carilli, J.E., 2009. Surface-temperature trends and variability in the low-latitude North Atlantic since 1552. *Nature Geoscience*, 2(7), pp.492–495.

Schlesinger, M.E. and Ramankutty, N., 1994. An oscillation in the global climate system of period 65–70 years. *Nature*, 367(6465), pp.723–726.

Siegel, P.E., Jones, J.G., Pearsall, D.M., Dunning, N.P., Farrell, P., Duncan, N.A., Curtis, J.H. and Singh, S.K., 2015. Paleoenvironmental evidence for first human colonization of the eastern Caribbean. *Quaternary Science Reviews*, *129*, pp.275–295.

Small, R.J.O., De Szoeke, S.P. and Xie, S.P., 2007. The Central American midsummer drought: Regional aspects and large-scale forcing. *Journal of Climate*, 20(19), pp.4853–4873.

Stephenson, T.S., Vincent, L.A., Allen, T., Van Meerbeeck, C.J., McLean, N., Peterson, T.C., Taylor, M.A., Aaron-Morrison, A.P., Auguste, T., Bernard, D. and Boekhoudt, J.R., 2014. Changes in extreme temperature and precipitation in the Caribbean region, 1961–2010. *International Journal of Climatology*, *34*(9), pp.2957–2971.

Stoddart, D.R., Bryan, G.W. and Gibbs, P.E., 1973. Inland mangroves and water chemistry, Barbuda, West Indies. *Journal of Natural History*, 7(1), pp.33–46.

Street-Perrott, F.A., Hales, P.E., Perrott, R.A., Fontes, J.C., Switsur, V.R. and Pearson, A., 1993. Late Quaternary palaeolimnology of a tropical marl lake: Wallywash Great Pond, Jamaica. *Journal of Paleolimnology*, 9(1), pp.3–22.

Taylor, M.A., Clarke, L.A., Centella, A., Bezanilla, A., Stephenson, T.S., Jones, J.J., Campbell, J.D., Vichot, A. and Charlery, J., 2018. Future Caribbean climates in a world of rising temperatures: The 1.5 vs 2.0 dilemma. *Journal of Climate*, 31(7), pp.2907–2926.

Taylor, M.A., Enfield, D.B. and Chen, A.A., 2002. Influence of the tropical Atlantic versus the tropical Pacific on Caribbean rainfall. *Journal of Geophysical Research: Oceans*, 107(C9), pp.10–11.

Taylor, M.A., Stephenson, T.S., Owino, A., Chen, A.A. and Campbell, J.D., 2011. Tropical gradient influences on Caribbean rainfall. *Journal of Geophysical Research: Atmospheres*, *116*(D21). https://agupubs.onlinelibrary.wiley.com/doi/full/10.1029/2010JD015580.

Tierney, J.E., Abram, N.J., Anchukaitis, K.J., Evans, M.N., Giry, C., Kilbourne, K.H., Saenger, C.P., Wu, H.C. and Zinke, J., 2015. Tropical sea surface temperatures for the past four centuries reconstructed from coral archives. *Paleoceanography*, *30*(3), pp.226–252.

Trouet, V., Harley, G.L. and Domínguez-Delmás, M., 2016. Shipwreck rates reveal Caribbean tropical cyclone response to past radiative forcing. *Proceedings of the National Academy of Sciences*, *113*(12), pp.3169–3174.

Vare, L.L., Massé, G., Gregory, T.R., Smart, C.W. and Belt, S.T., 2009. Sea ice variations in the central Canadian Arctic Archipelago during the Holocene. *Quaternary Science Reviews*, *28*(13–14), pp.1354–1366.

Vásquez-Bedoya, L.F., Cohen, A.L., Oppo, D.W. and Blanchon, P., 2012. Corals record persistent multidecadal SST variability in the Atlantic Warm Pool since 1775 AD. *Paleoceanography*, *27*(3). https://agupubs.onlinelibrary.wiley.com/doi/full/10.1029/2012PA002313.

Wang, C., 2007. Variability of the Caribbean low-level jet and its relations to climate. *Climate Dynamics*, 29(4), pp.411–422.

Wang, C., Lee, S.K. and Enfield, D.B., 2008. Atlantic warm pool acting as a link between Atlantic multidecadal oscillation and Atlantic tropical cyclone activity. *Geochemistry, Geophysics, Geosystems, 9*(5). https://agupubs.onlinelibrary. wiley.com/doi/full/10.1029/2007GC001809.

Wanner, H., Beer, J., Bütikofer, J., Crowley, T.J., Cubasch, U., Flückiger, J., Goosse, H., Grosjean, M., Joos, F., Kaplan, J.O. and Küttel, M., 2008. Mid-to Late Holocene climate change: An overview. *Quaternary Science Reviews, 27*(19–20), pp.1791–1828.

Watanabe, T., Winter, A. and Oba, T., 2001. Seasonal changes in sea surface temperature and salinity during the Little Ice Age in the Caribbean Sea deduced from Mg/Ca and 18O/16O ratios in corals. *Marine Geology, 173*(1–4), pp.21–35.

Wilson, S.M., 2007. *The Archaeology of the Caribbean.* New York: Cambridge University Press.

Winter, A., Ishioroshi, H., Watanabe, T., Oba, T. and Christy, J., 2000. Caribbean sea surface temperatures: Two-to-three degrees cooler than present during the Little Ice Age. *Geophysical Research Letters, 27*(20), pp.3365–3368.

Winter, A., Miller, T., Kushnir, Y., Sinha, A., Timmermann, A., Jury, M.R., Gallup, C., Cheng, H. and Edwards, R.L., 2011. Evidence for 800 years of North Atlantic multi-decadal variability from a Puerto Rican speleothem. *Earth and Planetary Science Letters, 308*(1–2), pp.23–28.

2 Water use and availability on Barbuda from the colonial times to the present

An intersection of natural and social systems

Rebecca Boger and Sophia Perdikaris

2.1 Introduction

Water has helped shape the history of the Barbuda landscape and the people living there. This chapter builds out of the initial work done by Boger et al. (2014) with new data; shows patterns that are unfolding about the physical characteristics, quality, and quantity of water available to Barbudans; and raises questions about how to leverage the aquifer system in Barbuda now and in the near future. This may be particularly important after extreme events like Hurricane Irma. The spread of disease and the lack of available clean water were given as the primary reasons by the Prime Minister of Antigua and Barbuda for the forced, military-led evacuation after Irma (Chappell 2017; Lyons 2017). After Irma, the Prime Minister led the repeal of the Barbuda Land Act (Antigua and Barbuda 2008) and opened the door to privatization of what was communally-owned land for tourist development, including housing, golf courses, and a marina (Perdikaris et al. 2021a, 2021b). The type of development and the amount of water used for these developments could have a significant impact on the aquifer system, including how water withdrawals change salinity and how contaminants enter the groundwater spread. Excavation for a large airport began shortly after Irma struck; work was conducted 24 hours a day and 7 days a week (Boger and Perdikaris 2019). Excavation has started or there are plans for one or two golf courses, gated communities, and a marina, which will all impact the island ecology as the landscape becomes 'Disneyfied' (Perdikaris et al. 2021). These changes are happening at a quick pace.

Although having enough good quality freshwater has always been a challenge, Barbuda fortunately does have an aquifer system that has done well to support people and wildlife on the island over the centuries. Barbudans in the colonial times developed strategies that leveraged the island's ecology, and people lived sustainably with great resilience (Boger et al. 2014, 2016; see Potter, Chapter 5 in this volume). They learned early on how to navigate the dry and wet annual seasons and longer-term droughts. Wildlife and livestock were and are still allowed to roam freely around the island to

DOI: 10.4324/9781003347996-3

eat and find sources of water in places during droughts (Berleant-Schiller 1983). Even now, Barbudans provide water to wildlife and roaming livestock during the dry periods through a network of historic wells (Boger et al. 2014). Whether or not Barbudans can weather the current socio-economic forces of disaster capitalism remains to be seen (Boger and Perdikaris 2019; Perdikaris et al. 2021a, 2021b; see Ibrahimpašić et al., Chapter 7 in this volume). Having better information on the groundwater system, in terms of both water quality and quantity available for use throughout the year, will be extremely important in informing decisions that can affect resilience and sustainability of Barbudans.

This chapter first starts with a brief overview of the climate, followed by a discussion of its karstic geology, which then provides context for the historical uses of water and the results of monitoring that has been undertaken between 2014 and 2019.

2.2 Climate

Barbuda has a tropical climate with dry and wet seasons. The driest months are January–March and the wettest months are August–November. The average temperature in Codrington is 26.7°C, while the average annual rainfall is 924 mm (Climate-Data.org). The Köppen and Geiger climate classification classifies Barbuda as an Aw, tropical savanna climate with non-seasonal or dry-winter characteristics. In addition to the seasonal dry months, Barbuda experiences recurrent periods of drought (Boger et al. 2014; Jackson 2001). The surrounding ocean waters average around 27°C, with little variation of around 3°C. The island is located in the northeasterly trade wind zone and experiences velocities ranging between 30 and 48 km/h. For more information on the climate, see Burn et al., Chapter 1 in this volume (Figure 2.1).

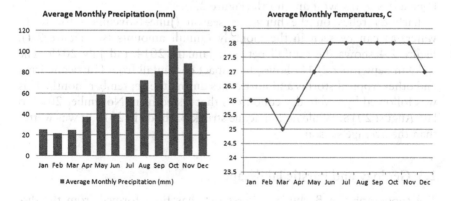

Figure 2.1 Average monthly precipitation and temperatures for Barbuda. Source: Climate-Data.org.

Table 2.1 Record of named hurricanes within 65 nautical miles of Barbuda

Name	Year	Dates	Name	Year	Dates
Alice	1954	Dec. 30–Jan. 6	Gaston	2010	Sept. 1–Sept.. 8
Arthur	1984	Aug. 28–Sept. 5	Georges	1998	Sept. 15–Oct. 1
Baker	1950	Aug. 18–Sept.1	Gonzalo	1984	Oct. 11–Oct. 20
Bertha	1996	Jul. 5–Jul. 17	Helena	1963	Oct. 25–Oct. 29
Betsy	1965	Aug. 27–Sept. 13	Helene	2000	Sept. 15–Sept. 25
Chris	2006	Aug. 1–Aug. 6	Ione	1955	Sept. 10–Sept. 27
Christine	1973	Aug. 25–Sept. 4	Irene	2011	Aug. 21–Aug. 30
Claudette	1979	Jul. 15–Jul. 29	Iris	1995	Aug. 22–Sept. 7
Daisy	1962	Sept. 29–Oct. 29	Irma	2017	Aug. 30–Sept. 13
Debby	2000	Aug. 19–Aug. 24	Jenny	1961	Nov. 1–Nov. 9
Dog	1950	Aug. 30–Sept. 18	Jose	1999	Oct. 17–Oct. 25
Donna	1960	Aug. 29–Sept. 14	Jose	2017	Sept 4–Sept. 25
Doria	1971	Aug. 20–Aug. 29	Klaus	1990	Oct. 3–Oct. 9
Earl	2010	Aug. 24–Sept. 6	Lenny	1999	Nov. 13–Nov. 23
Eloise	1975	Sept. 13–Sept. 24	Luis	1995	Aug. 27–Sept. 12
Faith	1966	Aug. 21–Sept. 7	Maria	2011	Sept. 6–Sept. 16
Fiona	2010	Aug. 30–Sept. 4	Noel	2007	Oct. 24–Nov. 6
Floyd	1981	Sept. 3–Sept. 12	Olga	2007	Dec. 10–Dec. 16
Frederic	1979	Aug. 29–Sept. 15			

Source: https://coast.noaa.gov/hurricanes/.

Barbuda is located within the Atlantic tropical hurricane belt. There are over 70 recorded storms and hurricanes that have passed within 65 nautical miles of Barbuda. In addition to the 37 listed in Table 2.1, there are 37 unnamed storms between 1852 and 1987 (NOAA). Most hurricanes occur between August and October, with September being the month with the most hurricanes (Figure 2.2).

On September 6, 2017, category 5 Hurricane Irma directly hit Barbuda and severely impacted the built and natural environments. Before and after satellite imagery shows how the vegetation was either killed or stripped of its leaves from the high winds or the storm surge that submerged the low-lying areas to the west and north (Figure 2.3).

Barbuda relies on the hurricane season (June–November) for freshwater, as can be seen in the monthly rainfall amounts (see Figure 2.1). Figure 2.4 shows the rainfall between January 2014 and July 2019. The spike in September 2017 shows the impact of rainfall from Hurricane Irma and other tropical storms and hurricanes during the September month. This was followed by severe dry conditions that lasted from November 2017 to late August 2018. While no hurricane struck Barbuda in 2018, it was wetter than the average season.

2.3 Geology and karstic landscape

The topography of Barbuda is karst that has been formed from the dissolution of the limestone bedrock that dominates the island. The tropical

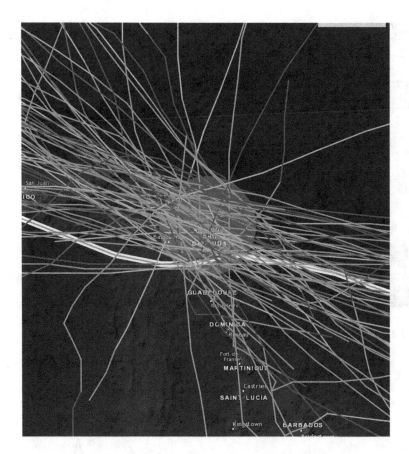

Figure 2.2 Hurricane tracks that have passed within 65 nautical miles of Barbuda. Source: https://coast.noaa.gov/hurricanes/.

climate with varying precipitation, interaction with saline ocean water, and changing sea levels over the millennia provided the conditions to create the landscape seen today. Barbuda's dominant feature is the Highlands, a raised plateau that reaches about 35 m above sea level. The Highlands occupies about one-fifth of the landmass; it is bounded by an abrupt escarpment along the east and north, and a gentler slope to the west and south.

Jagged limestone cliffs from the eastern Atlantic Ocean side along the Highlands. Exposed limestone pavements, stacks, and collapsed rocks can be seen along the coastline and in the interior. Caves and sinkholes typically found in karstic landscapes are common and have played an important role in Barbudan's cultural heritage (Perdikaris et al. 2013). Some caves or caverns, such as the Box Caves to the south have pools that are either fresh, brackish, or even saline in water quality. Subsurface caverns and tunnels are also common throughout much of the island and can be challenging for

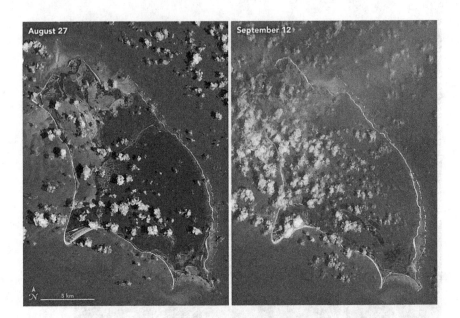

Figure 2.3 Comparison of satellite imagery taken before and after Hurricane Irma. The image to the right shows that lack of living vegetation with green foliage. Source: https://earthobservatory.nasa.gov/images/90975/barbuda-and-saint-barthelemy-browned-by-irma.

Rainfall (mm)

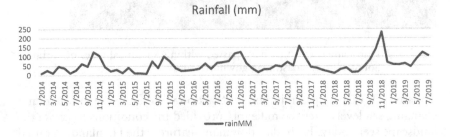

Figure 2.4 Rainfall between January 2014 and July 2019. Source: Climate-Data.org.

modern development. After Hurricane Irma, excavation for a large airport began; construction for the new airport eventually had to be abandoned after three separate attempts due to collapsing underground caves (Boger and Perdikaris 2019; Perdikaris et al. 2021b). Despite legal actions and local resistance to halt the airport construction (Barbuda Silent No More n.d.; The Daily Observer 2018; Garden Court Chambers 2017) and difficult geology for this type of endeavor, construction continues despite flooding and collapsing land that is forming new sinkholes (Perdikaris et al. 2021b).

This illustrates the importance of understanding the geology of the island when such large-scale developments are undertaken. An environmental impact assessment (EIA) was done after construction began and showed significant gaps in the EIA and 'failed to properly assess archaeology, biodiversity, hydrogeological and geological aspects' (Environmental Justice Atlas).

Brasier and Donahue (1985) and Brasier and Mather (1975) identify four geological formations with differing limestone characteristics. Figure 2.5 shows a map of the geology adapted from Brasier and Donahue (1985) and using Google Earth high-resolution imagery, digital elevation models, and field observations. The oldest mid-Miocene formation, the Highlands Formation, followed in age by the early Pleistocene Beazer Formation are limestones that have similar characteristics. Both are fine-grained and hard and originally were thought to be part of the same formation. The more recent, late Pleistocene Codrington Formation resulted from changing sea levels that rose from about 6 m above the present datum to below the present sea level.

> This resulted in the emergence of fringing and barrier reefs, the progradation of beach fringing and barrier reefs, the progradation of beach ridge and dune sands, and the formation of restricted lagoons, and salt ponds with algal flats.
>
> (Brasier and Donahue 1985, 1108)

The youngest Holocene formation, Palmetto Formation, is made from reefs, beaches, and lagoons building out from the older formations. Both the Codrington and Palmetto Formations are characterized by beach ridges and cheniers.

The karstic topography helps set the stage for hydrology. There are no permanent surface streams, and it has a semi-arid climate. While the hydrology has been divided into ten watersheds, these are poorly defined due to the low relief and frequent depressions in the karstic topography (Cooper 2001). The thin, often clayey soils and hardpan limestone contribute to rapid runoff and little retention of water through infiltration, and often it becomes difficult to transverse the landscape. Surface water collects in sinkholes and caves, limestone pavement depressions, and human features such as abandoned quarries used for limestone extraction and sand mining. These features play a critical role in the ecology. They form reservoirs for water during the rainy season, which are important for the wildlife during the dry seasons, especially during periods of drought. Bull Hole and Freshwater Ponds located central to the southwest of the island are brackish inland wetlands formed from flat smooth limestone pavement depression. This area is almost always with water, falling during droughts. Humanmade brackish ponds are located to the south near the former resorts of Coco Point and K Club. Coastal hypersaline ponds are found to the south and southwest and are not hydraulically connected to the ocean, whereas Hogs Pond along the northeast is an anchialine pond which is a collapsed sinkhole connected to the ocean and fed by seawater (Figure 2.6).

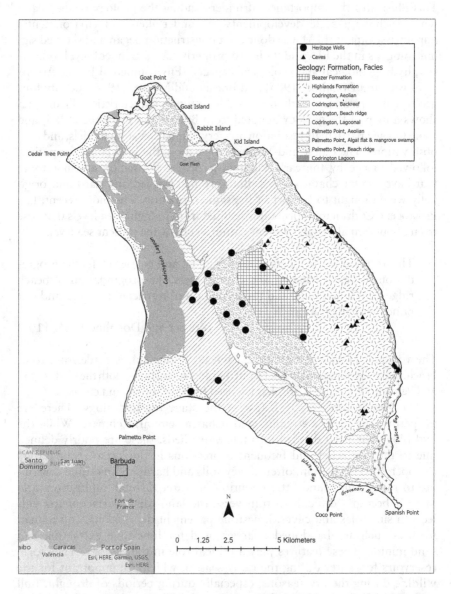

Figure 2.5 Map of historic wells, caves, and geology of Barbuda. Geology based on Brasier and Donahue (1985), satellite imagery, and field observations. Map created by Rebecca Boger.

2.4 Historic wells

Despite the fact that there are no permanent streams and only limited surface water, there is an extensive aquifer system that has been utilized by people living there for centuries. During the Colonial period, a network of historic wells were dug in varying locations, largely in the lower-lying areas

Figure 2.6 Aerial photo of Beazer Well showing the general layout of many of the historic wells. Photo taken by Rebecca Boger.

to the west, spreading out from Codrington and reaching to the northeast and to the south (see Figure 2.7). These wells are in varying stages of preservation. The wells and their associated structures are based on a similar design. Each well has a trough attached to it from which the animals drank. The well and trough are surrounded by a stone wall with a gate. Some have multiple troughs at different heights to accommodate for the differential size of the livestock (e.g., cattle, goats, sheep). Within the wall, there is an enclosed pen (Sluyter 2012). Figure 2.6 shows an aerial photo of Beazer well as an example of a general layout design. The truck near the top of the photo gives a sense of scale.

Barbuda, unlike neighboring Antigua and many other Caribbean islands, did not undertake large-scale agriculture production. Its thin soils and arid climate were not conducive to sugarcane, cotton, and other cash crops (Harris 1965; Berleant- Schiller 1977, Boger et al. 2014). Barbuda experiences long winter-dry season and frequent droughts (Boger et al. 2014). To successfully live within these conditions, Barbudans developed sustainable practices where they essentially abandoned commercial cultivation during the dry years and expanded stock (e.g., cattle) production. During wet years and seasons, the animals would be able to roam and drink freely from the

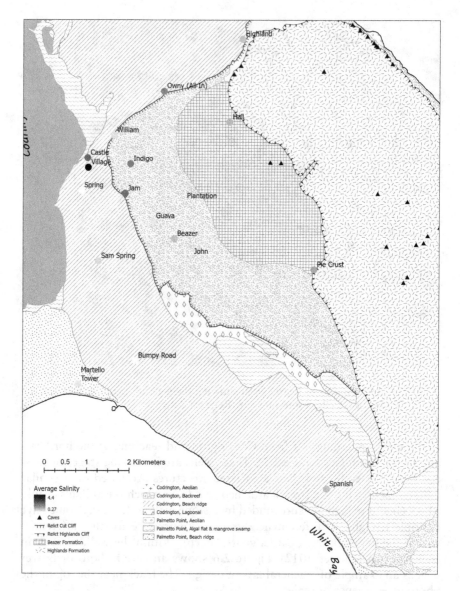

Figure 2.7 Map showing the geology, location, and average salinity values of the wells sampled. Wells shown in black are not regularly sampled.

troughs and natural sinkholes and other karstic formations where water would gather. When needed, Barbudans would round them as needed for butchering and export. During wet years, crop production would increase. During dry years and seasons, the livestock would gather at the wells for water. Once there, Barbudans would confine the livestock in the pens for culling. These practices leveraged the ecology of the island and were made possible by the communal ownership of land 'Land use and tenure together have preserved the community from the hazards of drought and domination' (Berleant-Schiller, 1983, 87).

2.5 Water quality of wells

Between 2014 and 2017, the water in the wells has been tested for salinity using a YSI® probe twice a year, in January and late May/early June. These months coincide toward the ends of dry and wet seasons. In wells with water close to the surface, the probe was placed in the well directly for the readings. Where the water was too deep for the cable, water samples were taken with a bucket. The bucket was rinsed thoroughly before taking readings with a new sample to make sure that the values correctly documented the well water. In addition, Solinst© conductivity and barometric probes were placed in the well on the grounds of the Sir McChesney George Secondary School. This well is not a historic well and is located near the Indigo well. These probes collected conductivity and air pressure values every hour. In January 2018, these probes were removed from the Secondary school well and placed in the Sam Spring well since the former well was filled with debris from Hurricane Irma in September 2017. In 2018, the YSI meter and probe were replaced using a Hanna® meter and probe. In 2016, the well water was tested for fecal coliform using a LaMotte® test kit. All wells tested positive for fecal coliform.

Table 2.2 lists the average salinity and standard deviation (SD) values for the water in the historic well water. The table also shows the number of times the wells were sampled. Missing values were caused by several reasons such as the inability to access roads due to flooding or excessive tree debris, or the wells became full of garbage and other debris so that the probe could not be adequately immersed and no water could be collected with a bucket. Monitoring shows that Village well has a much higher salinity than the other wells and is more than 2 ppt higher than that of Castle, the well with the next highest salinity at 2.24 ppt. Spring and Bumpy wells have the lowest salinity values at 0.27 ppt and 0.29 ppt, respectively, while the others range from 0.46 to 2.24 ppt.

Figure 2.7 shows a map of the average salinity values for well water and the underlying geology adapted from Brasier and Donahue (1985) for the formations and facies where the wells are located. Wells with higher salinity values are located within or near where most people live in Codrington, with the exception of William and Spring wells that have some of the lowest salinity values although located nearby the others mentioned.

Several wells are located within or close to the contact zones of the geological units. Hall and Pie Crust wells are located on the contact between Beazer and Highland Formations; Jam and William wells are on the zone between beach ridge facies of the Codrington Formation; and Owney is on the contact between the beach ridge and backreef facies of the Codrington

Table 2.2 Average salinity, standard deviation (SD), and the number of times tested for the historic wells between 2014 and 2019

Well	Average salinity (ppt)	SD	Number of times tested
Spring	0.27	0.07	9
Bumpy	0.29	0.19	10
Guava	0.46	0.25	6[a]
William	0.57	0.13	10
Sam Spring	0.69	0.51	10
Martello[b]	1.62 (0.54 before Irma)	1.98 (0.19 before Irma)	9
Highland	0.94	0.21	10
Beazer	0.96	0.31	9
Spanish	1.04	0.11	7
Hall	1.06	0.11	6
Pie Crust	1.27	0.24	7
Owney	1.74	0.48	10
Indigo	1.83	0.25	8
Jam	2.10	0.62	10
Castle	2.24	0.79	9
Village	4.40	0.48	8

Note: The wells are ranked in order of lowest average salinity to highest and color coded into three categories (high, medium, and low).

[a] Started testing this well in May 2016 after learning about its existence.
[b] Martello Tower included in the wells with lower salinity. It was greatly affected by the storm surge from Hurricane Irma.

Formation. Most of the other wells fall within the beach ridge and backreef facies of the Codrington Formation. Lastly, Martello Tower well is within the Palmetto Formation. Although located very close to the open Caribbean Sea, the water in the Martello Tower well has very low salinity. It was, however, the most impacted by Hurricane Irma since the Palmetto area was inundated by the storm surge.

Figure 2.8(a)–(c) shows the graphs of the well water between 2014 and 2019. For ease of interpretation, the 16 wells are separated into three graphs based on their relative salinity (high, mid, and low). Many of the wells show a seasonal cycle of higher and lower salinity values in January and May/June, respectively, corresponding to the ends of the wet and dry seasons. The pattern is the strongest in the wells grouped with mid-range values. Sam Spring well shows a different pattern between 2016 and 2018 when it increases in salinity up to 1.7 ppt and then drops.

In looking at the data logger data from May 2016 to January 2018 for the Secondary school well located near Indigo well (Figure 2.9(a)), one can see the seasonal variability of salinity values with higher salinity values between July and October 2016 and lower salinity values between November and May, and then starts to rise until September 2017. There are two sharp declines: one in November 2016 and another in September 2017.

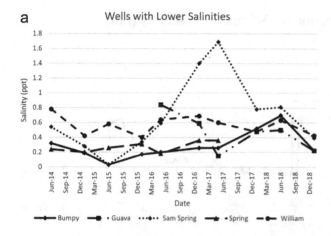

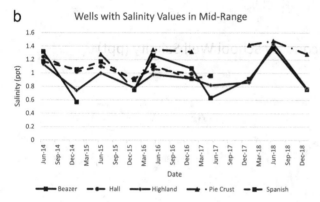

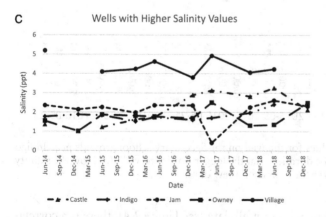

Figure 2.8 (a)–(c) Graphs showing salinity between 2014 and 2019. Note that there are different scales on the y-axis.

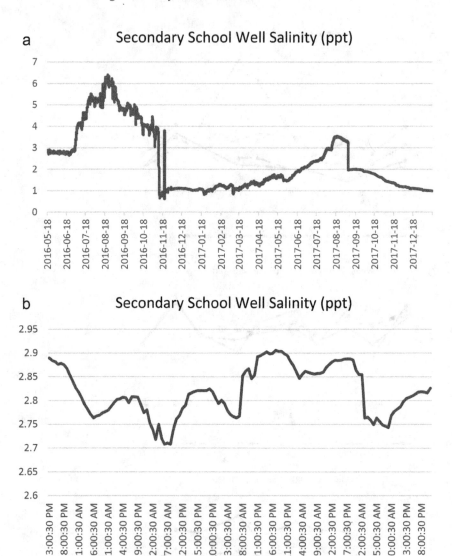

Figure 2.9 Salinity values for the well on the Secondary school grounds near Indigo
well. (a) The duration that the logger was in the well. (b) How the salinity
changes over the course of a few days.

Comparing the decline with rainfall data (see Figure 2.4), there is a correlation between increased rainfall at that time and decreased salinity values in the well water. The September 2017 drop coincides with Hurricane Irma. Figure 2.9(b) shows more temporal detail for the salinity values between May 18 and 23, 2018. There are daily fluctuations of higher and lower salinity values that likely result from tidal fluctuations.

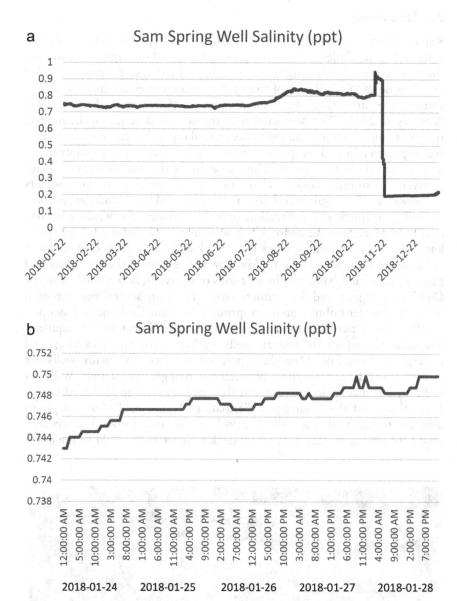

Figure 2.10 Salinity values for Sam Spring well. (a) The duration that the logger was in the well (b) How the salinity changes over the course of a few days.

The water at Sam Spring well shows a different pattern from January 2018 to January 2019 (Figure 2.10(a) and (b)). The salinity values are fairly stable at 0.75 ppt until October 2018 when it drops quickly to about 0.2 ppt. The drop corresponds to increased rainfall shown in Figure 2.4. Likewise, the daily fluctuations are not so large, less than 0.002 ppt versus 0.1 ppt seen at the Secondary school well.

2.6 Discussion

Karstic geology is very variable. The surface landscape and below are characterized by sinkholes, caves, springs, conduits, and underground drainage that are formed by dissolution, drainage, and collapse processes (Taylor and Greene 2008). Because of their high degree of heterogeneity and anisotropy, they are difficult to model numerically (Ghasemizadeh et al. 2012; Wang et al. 2016). The permeability of the aquifer depends on the matrix, fractures, and conduits, and the underground pipes that carry water from a recharge area to an outlet spring. Groundwater storage occurs mainly in the matrix, and the matrix permeability determines the transmission of water between the storage blocks and the porosity of the fractures or conduits (Atkinson 1977). Aquifer recharge is through sinkholes and depressions, as well as through soil infiltration into the epikarst, the weathered zone near the surface, into the bedrock below (White 1999); most discharge from karst systems is through springs (Ford and Williams 2013). Groundwater in limestone islands will tend to flow downslope toward the sea (Jones and Banner 2003), and based on the locations of the wells, flow would be to the Codrington lagoon and the southern coast. There are several sites reported where freshwater bubbles up from springs within the Codrington lagoon.[1]

The wells in Barbuda are tapping into unconfined near-surface aquifers. The sites selected for the historic wells are likely tapping into springs and underground conduits from these perched aquifers, and many wells are located along the contact zones between different geological units. Williams well, for instance, is located right at the line between the Highland/Beazer Formations and the Codrington Formation. Figure 2.11 shows a photo adjacent to the Williams well. The fork in the road shows a sharp contrast between the surface soils where dark soils are to the right to the north

Figure 2.11 Fork in the road adjacent to William well, located to the left of the car.

while the lighter color soils are to the left above the backreef facies of the Codrington Formation.

The groundwater exists as a freshwater lens with estimated thicknesses ranging from little to none in the Highland Formation to 4 m in the Beazer Formation and 10 m in the Palmetto Point Formation. The thickness of the lens is associated with the age of the limestone, with the thickest lens in the unconsolidated sands of the Palmetto Point Formation and the thinner lenses found with greater diagenesis lithification flow occurring more in fractures (Mather 1988). The Palmetto Point aquifer has been compromised by sand mining for many years and potentially more recently with the Peace, Love, and Happiness development. The removal of sand has raised the water table and exposed part of the aquifer to drying out (Cooper and Bowen 2001).

Transmissivity is the rate at which water flows through the saturated thickness of an aquifer and depends on the permeability structure such as the shape, amount, and interconnectivity of the void spaces (Galvão et al. 2017). The transmissivity of the aquifers will depend on the karstic system behavior at any particular location. Those wells tapping into high transmissivity will recharge the well more rapidly with withdrawals. In speaking with local residents, the wells known to recharge quickly are William, Beazer, and Jam. In fact, Jam well is used extensively by residents to fill cisterns for human uses and water livestock. Jam, William, and Owney are located along the relict cliff between the Codrington beach ridge and Codrington backreef facies (see Figure 2.7).

The water in the wells is also susceptible to saline incursion during periods of drought and during storm events, where surges and swells force saline water within the shallow saturated limestone aquifer. The wells with the higher salinities (Castle, Village, Owney, Indigo, Jam, and Secondary school) are located within or near Codrington where most people live. These wells appear to be more susceptible to saltwater intrusion, perhaps resulting from greater extraction of water from the wells and subsequent depression for the saline water to flow into. The hourly data from the well at the Secondary school indicates greater daily salinity fluctuations than found at Sam Spring (although the datasets are taken at different time periods). This could indicate that the thin perched aquifer lens is subjected to mixing of the fresher lens and more saline water under it through tidal activity. The daily fluctuations of salinity, although minor, in Sam Spring indicate that the conduit system may be greater here than at the Secondary school well and that the greater freshwater input masks the tidal signature.

More research needs to be done to understand the water resources for human use, in particular, how much can be withdrawn without changing the salinity or impairing its ability to recharge. As well, more monitoring is needed to better understand how the aquifers are responding to climate change, in particular the annual amount and seasonal distribution of precipitation and rising sea levels. If extreme events, both drought and hurricanes, increase (Cubasch et al. 2013; Nurse et al. 2014), then there could be greater seasonal salinity fluctuations. With rising sea levels, the salinity may become intensified, particularly during the dry seasons. Caribbean karst landscapes can be particularly sensitive to the impacts of climate change and human activity (Day 1993), and this can result in intensified challenges. Barbudans in the past have navigated the challenges successfully, and this knowledge has been passed through the generations. A challenge for Barbudans today is how to integrate scientific monitoring and modeling with the traditional knowledge for sustainable and resilient water practices.

Note

1 Based on communication with residents.

References

Antigua and Barbuda, 2008. The Barbuda Land Act 2007. *Official Gaz*, 28(5): 1–18. Available at faolex.fao.org/docs/pdf/ant78070.pdf.

Atkinson, T.C., 1977. Diffuse flow and conduit flow in limestone terrain in the Mendip Hills, Somerset (Great Britain). *Journal of Hydrology*, 35(1–2), pp.93–110.

Barbuda Silent No More. (n.d.a). Barbuda silent no more. *Facebook*. https://www.facebook.com/barbudasilentnomore.

Berleant-Schiller, R., 1977. The social and economic role of cattle in Barbuda. *Geographical Review*, 67, pp.299–309.

Berleant-Schiller, R., 1983. Grazing and gardens in Barbuda. In R. Berleant-Schiller & E. Shanklin (Eds.),*The Keeping of Animals: Adaptation and Social Relations in Livestock Producing Communities*. Totowa, N.J.: Allanheld, Osmun pp.73–91.

Boger, R. and Perdikaris S., 2019. After Irma, disaster capitalism threatens cultural heritage in Barbuda, North American Caribbean and Latin America (NACLA), Feb. 11, 2019. Available at https://nacla.org/author/Rebecca%20Boger%20and%20Sophia%20Perdikaris.

Boger, R., Perdikaris, S., Potter, A.E. and Adams, J., 2016. Sustainable resilience in Barbuda: Learning from the past and developing strategies for the future. *International Journal of Environmental Sustainability*, 12(4), pp.1–14.

Boger, R., Perdikaris, S., Potter, A.E., Mussington, J., Murphy, R., Thomas, L., Gore, C. and Finch, D., 2014. Water resources and the historic wells of Barbuda: Tradition, heritage and hope for a sustainable future. *Island Studies Journal*, 9(2), pp.327–342.

Boger, R., Perdikaris, S. and Rivera-Collazo, S.P.I., 2019. Cultural heritage and local ecological knowledge under threat: Two Caribbean examples from Barbuda and Puerto Rico. *Journal of Anthropology and Archaeology*, 7(2), pp.1–14.

Brasier, M.D. and Mather, J.D., 1975. The stratigraphy of Barbuda, West Indies. *Geological Magazine*, 112(3), pp.271–282.

Brasier, M. and Donahue, J., 1985. Barbuda—an emerging reef and lagoon complex on the edge of the Lesser Antilles island are. *Journal of the Geological Society*, 142(6), pp.1101–1117.

Burn, M.J., Boger, R., Holmes, J., and Bain, A. 2022. A long-term perspective of climate change in the Caribbean and its impacts on the island of Barbuda. In *Barbuda: Changing Times, Changing Tides.* Chapter 1. This volume

Chappell, B. 2017. 3 weeks after Irma wrecked Barbuda, island lifts mandatory evacuation order. *NPR.* [Online] Available: https://www.npr.org/sections/thetwo -way/2017/09/29/554540066/3-weeks-after-irmawrecked-barbuda-island-lifts -mandatory-evacuation-order (September 29, 2017).

Climate-Data.org. https://en.climate-data.org/north-america/antigua-and-barbuda -165/#example0.

Cooper, B. and Bowen, V., 2001. *Integrating Management of Watersheds & Coastal Areas in Small Island Developing States of the Caribbean: National Report for Antigua & Barbuda.* Ministry of Tourism and Environment.

Cooper, V. 2001. *Inland Flood Hazard Mapping for Antigua and Barbuda: Summary Report.* OAS/USAID. http://www.oas.org/pgdm/hazmap/flood/abfldsum.htm.

Cubasch, Ulrich, Wuebbles, Donald, Chen, Deliang, Facchini, Maria Cristina, Frame, David, Mahowald, Natalie and Winther, Jan-Gunnar., 2013. Introduction in Climate Change 2013. In T.F. Stocker, D. Qin, G.-K. Plattner, M. Tignor, S.K. Allen, J. Boschung, A. Nauels, Y. Xia, V. Bex & P.M. Midgley (Eds.), *Intergovernmental Panel on Climate Change 2013: The Physical Science Basis. Contribution of Working Group I to the Fifth Assessment Report of the Intergovernmental Panel on Climate Change,* Cambridge, United Kingdom and New York, NY, USA: Cambridge University Press. pp.119–158.

The Daily Observer. (2018, September 12). Barbuda airport injunction lifted. *The Daily Observer.* https://antiguaobserver.com/barbuda-airport-injunction-lifted/.

Day, M.J., 1993. Human impacts on Caribbean and Central American karst. *Catena. Supplement (Giessen), 25,* pp.109–125.

Environmental Justice Atlas. https://ejatlas.org/print/barbuda-new-airport.

Ford, D. and Williams, P.D., 2013. *Karst Hydrogeology and Geomorphology.* John Wiley & Sons.

Galvão, P., Halihan, T. and Hirata, R., 2017. Transmissivity of aquifer by capture zone method: An application in the Sete Lagoas Karst Aquifer, MG, Brazil. *Anais da Academia Brasileira de Ciências, 89*(1), pp.91–102.

Garden Court Chambers. 2017. Barbudans file injunction to expose unlawful introduction of land act changes. Garden Court Chambers. https://www .gardencourtchambers.co.uk/news/barbudans-file-injunction-to-expose-unlawful -introduction-of-land-act-changes.

Ghasemizadeh, R., Hellweger, F., Butscher, C., Padilla, I., Vesper, D., Field, M. and Alshawabkeh, A., 2012. Groundwater flow and transport modeling of karst aquifers, with particular reference to the North Coast Limestone aquifer system of Puerto Rico. *Hydrogeology Journal, 20*(8), pp.1441–1461.

Harris, D.R., 1965. *Plants, Animals, and Man in the Outer Leeward Islands, West Indies; An Ecological Study of Antigua, Barbuda, and Anguilla,* Berkeley: University of California Press.

Ibrahimpašić, E., Perdikaris, S. and Boger, R. 2022. Disaster capitalism: Who has the right to control their future? In *Barbuda: Changing Times, Changing Tides.* Chapter 7. This volume

Jackson, I., 2001. *Drought Hazard Assessment and Mapping for Antigua and Barbuda: Post-Georges Disaster Mitigation Project in Antigua & Barbuda and St. Kitts & Nevis, April 2001.* Organization of American States, Unit for

Sustainable Development and Environment. https://www.oas.org/pgdm/hazmap/drought/abdrttec.pdf

Jones, I.C. and Banner, J.L., 2003. Estimating recharge thresholds in tropical karst island aquifers: Barbados, Puerto Rico and Guam. *Journal of Hydrology*, 278(1–4), pp.131–143.

Lyons, K., 2017. The night Barbuda died: How Hurricane Irma created a Caribbean ghost town. [Online] Available: https://www.theguardian.com/global-development/2017/nov/20/the-night-barbuda-diedhow-hurricane-irma-created-a-caribbean-ghost-town (November 20, 2017).

Mather, J.D., 1988. The influence of geology and karst development on the formation of freshwater lenses on small limestone islands. *Karst Hydrogeol. Karst Environ. Prot*, 1, pp.423–428.

NASA. https://earthobservatory.nasa.gov/images/90975/barbuda-and-saint-barthelemy-browned-by-irma.

NOAA. https://coast.noaa.gov/hurricanes/.

Nurse, L., McLean, R., Agard, J., Briguglio, L.P., Duvat, V., Pelesikoti, N., Tompkins, E. and Webb, A., 2014. *Climate Change 2014: Impacts, Adaptation and Vulnerability. Contribution of Working Group II to the Fifth Assessment Report of the Intergovernmental Panel on Climate Change.* Cambridge University Press, Cambridge, United Kingdom and New York, NY, USA.

Perdikaris, S., Boger, R., Gonzalez, E., Ibrahimpašić, E., and Adams, J., 2021a. Disrupted identities and forced nomads: A post-disaster legacy of neocolonialism in the island of Barbuda, Lesser Antilles. *Island Studies Journal*, 16(1):pp. 115–134.

Perdikaris, S., Boger, R. and Ibrahimpašić, E., 2021b. Seduction, promises and the disneyfication of Barbuda post Irma. *TRANSLOCAL Contemporary Local and Urban Cultures Journal.* Number 5 (un)inhabited spaces. Ana Salgueiro and Nuno Marques (eds). ISSN 2184-1519 Madeira and Umeå. Translocal.cm-funchal.pt/2019/05/02/revista05/

Perdikaris, S., Grouard, S., Hambrecht, G., Hicks, M., Mebane Cruz, A. and Peraud, R., 2013. The caves of Barbuda's eastern coast: Long term occupation, ethnohistory and ritual. *Caribbean Connections*, 3, pp.1–9.

Potter, A. 2022. From the far ground to the near ground: Barbuda's shifting agricultural practices. In *Barbuda: Changing Times, Changing Tides.* Chapter 5. This volume

Sluyter, A., 2012. *Black Ranching Frontiers: African Cattle Herders of the Atlantic World*, 1500–1900. Yale University Press. https://doi.org/10.12987/yale/9780300179927.003.0004

Taylor, C.J. and Greene, E.A., 2008. Hydrogeologic characterization and methods used in the investigation of karst hydrology. In *Field Techniques for Estimating Water Fluxes between Surface Water and Ground Water*, edited by D.O. Rosenberry and J.W. LaBaugh, pp.71–114. US Geological Survey, Reston, Virginia (EUA).

Wang, X., Jardani, A., Jourde, H., Lonergan, L., Cosgrove, J., Gosselin, O. and Massonnat, G., 2016. Characterisation of the transmissivity field of a fractured and karstic aquifer, Southern France. *Advances in Water Resources*, 87, pp.106–121.

White, W.B., 1999. Conceptual models for karstic aquifers. *Karst modeling*, 5, pp.11–16.

3 Developing agency and resilience in the face of climate change

Ways of knowing, feeling, and practicing through art and science

Jennifer D. Adams and Noel Hefele

There could be no creativity without the curiosity that moves us and sets us patiently impatient before a world that we did not make, to add it to something of our own making. (Freire 1998, 18)

3.1 Introduction

In every generation, young people have had to prepare for an uncertain future. Different eras saw different challenges, and young people have had to enact their daily lives while projecting future selves in a world that is only imagined to them, most often not with rose-colored lens. During the last century, Western cultures that produce neoliberal affluence have also caused unimaginable threats to our global survival. Climate change, food security, and loss of human and animal traditional places are some of the pressing issues that our young people will encounter and have to work to resolve. Effects of climate change are certain in some regards while uncertain in others. It is in this uncertainty that much of the progress towards global sustainability exists. Responding to climate change, in many respects, can be thought of as improvisational, with key components of improvisation involving responsiveness to the 'unforeseen' and the 'open-ended' (Douglas 2012). A habit or method of improvisation, which Douglas (2012) paraphrases Ingold's definition as 'a perpetual state of responsiveness through movement within a constantly shifting world' (para. 24), is a worthwhile strategy of engagement in an uncertain world. Climate change presents many challenges, known and unknown, that we, as a global collective, will have to respond to with the young people increasingly taking lead. How will we work with our youth to prepare them to face such challenges? How will we ensure that all youth, regardless of race, ethnicity, or socioeconomic status, have opportunities to engage in discussions about issues that will affect the quality of their lives? This is especially critical for those youth who continue to be systematically racialized and marginalized; how will we create experiences to ensure that they are at the proverbial table when it comes to making decisions that affect their individual well-being and that of their

DOI: 10.4324/9781003347996-4

communities? In order to do so, we need to create spaces where young people are able to make salient connections to place and access their creativity in ways that allow for the (re)imagination of sustainable, resilient, ethical, and productive futures for themselves and communities.

In this chapter, we will describe an engaged artist residency and participatory art project with youth on the island of Barbuda. We address the central question of 'how can art and science be integrated in working with youth in ways for them to learn about place, identity and connections to the natural world?' Our conceptualization of this project builds from de Sousa Santos' (2007) notion of ecologies of knowledge, and that knowledge exists only in the plurality of ways of knowing. In other words, knowledge does not exist in the silos that structure many of our academic institutions; however, it exists in the complex lived experiences of the interactions between people and the natural and built environments. This phenomenon is evidenced in many of the pressing issues of our times. Issues such as climate change, food security, and environmental sustainability do not only exist in the geoscience, biology, and chemistry content silos, as how it is taught in formal science, technology, engineering, and mathematics (STEM) education. Rather, these issues have transdisciplinary origins and require acknowledgment of the social, political, and cultural factors that all contribute to these issues. Disciplines, while recognizing their areas of expertise, will have to rely on transdisciplinary approaches to address these issues. Like scientists, artists have historically addressed compelling questions of our time (Adams and McCullough, in press). Artists and scientists have been engaged in explorations focused on the dynamic relationship among physical, cultural, and natural land, sea, and urbanscapes. Both science and art can be viewed as cultural tools in the service of addressing societal issues.

3.2 Part 1: learning with climate change

3.2.1 Critical imagining

Barbuda is facing unprecedented climate challenges for which the young people will have to make critical decisions about their livelihoods on the island – the culture will either have to adapt or perish; the stakes are that high. In order for youth to adapt, it is important that we develop conceptual spaces where young people could both solidify and expand their identities in connection to place while leveraging their existing knowledge and what they will learn towards imagining sustainable and resilient futures. Art allows for new ways in and through scientific thought and opportunities for students to try out new ways of relating to their world. The arts allow students to develop an expanded lens with which to view their worlds, new languages of expression and communication, and human-centered ways of engaging in science and the environment. We believe that integrated art and science skill-building can help develop a sense of agency and resilience in the face of climate change. As such we designed and initiated an engaged artist residency where a visual artist worked with the local youth to co-create artistic expressions that describe their identity as Barbudans in connection to their natural environment and the challenges they face within the specter of climate change. We also felt that with minimal opportunity to participate in art-related pursuits in school, the art/science engagement would afford the young people a new and collectively meaningful way to connect with their environment.

3.2.2 *Learning and grasping complex issues*

Climate change is an exceedingly difficult problem to grasp. We are bombarded by 'facts' that propel us into stupefied guilt and helpless inaction. While we become increasingly aware of the causes (i.e., increased carbon in the atmosphere) and potential effects (i.e., sea level rise), we often feel disempowered to make changes toward sustainability and resilience; our participation and individual contributions to it seem so small and insignificant. There are obvious inequities – the United States and other developed nations produce much of the world's greenhouse gases relative to developing nations, yet it is the developing nations who are most vulnerable to the effects of climate change for a combination of geographic and economic reasons (Levy and Patz 2015). Places like Barbuda are on the front lines of the impact of climate change, yet are statistically non-actors in terms of the causes (Adams et al. 2017). With the 2017 Hurricane Irma that ravaged the island, Barbuda is already experiencing extreme weather events that result from a changing climate. The young people will have to manage the repercussions of climate change and are left to ask the question, 'what kind of world do we want to live in in the future?' In the face of these seemingly grim prospects lies an opportunity for creativity in imagining sustainable, resilient, and just futures.

A starting point with the students felt like establishing the fact that their artwork would exist in the world, and that they could make artwork about their existence in the world. It felt important to ask questions to get students thinking about relationships and contexts, and that those could be subjects of their work, or the source of new questions, almost like a dialogue with the environment. 'To correspond with the world, in short, is not to describe it, or to represent it, but to *answer to it*' (Ingold 2013, 108); both art and science can be thought of in this way, but we must get past thinking of art as representational ideals and science as objective and static conveyors of truth. Art and science explore the world, thus creating spaces for truth and expressions of value. In other words, both science and art are active life paths of developing skilled habits (Ingold 2018), learning through movement through the world while shaping the future of our experience of being in the world, which Ingold calls an 'education of attention'. The nexus of art and science provides a salient space for young people to develop habits of creative correspondence and responsiveness to a rapidly changing world.

3.2.3 Approaching art and science

When speaking of the history of Barbuda, we often heard Barbudans describe their island as the 'breadbasket of Antigua' as it supplied much of the produce that was sold on the larger and more hegemonic sister island. For numerous reasons, local agriculture declined and the island became reliant on imported goods. Thus, the barge becomes a symbol of resources; when the barge does not travel due to weather or mechanical issues, Barbuda is without food and supplies. With increasing and severe weather incidents occurring, this barrier to food will become more common, rendering the populace food insecure. As such, the principal of Sir McChesney George Secondary School embarked on a mission to re-emphasize agriculture to both bring back the industry as a source of livelihood and contribute toward a more self-sufficient economy. This school leader is also a marine biologist and imparts both his western scientific and local knowledge to his students so that they have a better ecological understanding of their island home.

The engaged artist residency provided the opportunity to leverage the history and culture of Barbuda as well as foster the future thinking of students to imagine and create their place and impact on the land. The artist both pursues their own work while contributing to advancing the arts in the community. Barbuda offers much natural, cultural, and historical inspiration for an artist. The context also affords ample sources to elicit the innate creativity of the young people who live on the island.

We used the engaged art and science approach both as a way to learn about Barbudan youth identity in the face of climate change and as a way of engaging the young people in collaborative art-making. Rather than interviewing students about their experiences and prospects for the future, the arts engagement offered a way for us to learn about their experiences and identities while allowing the young people to develop a sense of creative agency that could extend towards imagining and realizing alternative futures for themselves and for their island.

3.2.4 Agency and resilience

Because of the unprecedented effects of climate change, we need to think creatively about notions of resilience in how people adapt to a shifting planet, especially the most vulnerable. For this work, we borrow a definition of resilience as proposed by the Arts Council, UK,

> By resilience we mean the vision and capacity of organisations to anticipate and adapt to economic, environmental and social change by seizing opportunities, identifying and mitigating risks, and deploying resources

effectively in order to continue delivering quality work in line with their mission.

(Woodley et al. 2018, 8)

It is maintaining a space of possibilities while not feeling overwhelmed by a problem at hand, and seeing and being able to accomplish a focused set of actions in line with a goal. A report commissioned by the Arts Council England delineates three distinct areas of resilience:

- **Resilience as 'bounce back' from shocks:** rebounding as quickly as possible to a previous state, with the implicit assumption that this was a stable state
- **Resilience as 'ability to absorb' shocks:** with a focus on maintaining the same 'structure, function and identity' in the face of shocks; 'the capacity of a system to absorb disturbance and still retain its basic function and structure'
- **Resilience as 'positive adaptability'** in anticipation of, or in response to, **shocks:** a system adapting its structure, functions, and operations in the face of new conditions

(Woodley et al. 2018, 9)

Furthermore, Woodley et al. (2018) distinguish between the first two definitions and the third. 'Bouncing back' refers to surviving, enduring, strength, returning to a prior state, and preserving the core mission and goals. This, in contrast to the latter, 'bouncing forward', which means thriving, evolving, flexibility and adaptation, changing, and developing mission and goals in light of changed circumstances and needs. The third meaning of resilience is more pertinent to our project as it demonstrates a sense of agency, which we believe works hand in hand with resilience, anticipating the world as a way to get ahead of a series of reactive responses.

We define the agency as the ability to use resources available in a given context in order to transform lived experiences (Adams and Gupta 2017). While it is a hierarchical distribution of skills, there is a question of how can those skills and voices be accessed and amplified in a given environment with concrete opportunities and constraints. Agency is an ongoing dialogue with the environment or context, as Biesta et al. (2017, 38–40) note,

> Against the tendency to think of agency as a capacity or ability individuals possess, we have pursued an ecological understanding of agency that focuses on the question of how agency is achieved in concrete settings and under particular ecological conditions and circumstances (Biesta and Tedder 2006). This ecological view of agency sees agency as an emergent phenomenon of the ecological conditions through which it is enacted.

Ingold (2018) calls this agency, not given in advance but the task and environment emergent, as *agencement*; it is 'a task we are bound to take on as responsive and responsible brings, and as part of the life we undergo' (24). For us, resilience and creative agency are almost two sides of the same coin, a stance to take to the world as you move through it and a framework that allows us to think about science and art-making as tools or avenues that allow young people to learn more about their environment with the goal of being able to remake their environment in the face of climate change. With creative agency, young people access physical and conceptual resources in their environments and transform them as they imagine resilient futures for themselves and their context.

3.2.5 Integrating art and science toward creative agency

All science depends on observation, and observation depends in turn on an intimate coupling, in perception and action, of the observer with those aspects of the world that are the focus of attention.

(Ingold 2011, 75)

While science has been long seen as critical in educating young people for productive futures, art has been marginalized as not important, especially for schools and students with limited resources. However, as Biesta (2017, 12) notes, 'the educational significance of the arts, and perhaps the educational urgency of the arts, lies in art education *beyond* expressivism and creativity'. Biesta (2017) goes on to explain that there is a third aspect of art in education. It is not the *instrumental justifications* where engagement of the arts is useful for its impact or significance on something else. It is not the flipside of instrumentalization of art, where it is merely art for art's sake, where art is 'fundamentally useless … without any value beyond art.' As Biesta says, 'expression in itself is never enough' (14). Biesta takes a 'world-centered' approach where the student is to exist in and with the world as a subject. He terms this as existing in a 'grown up' way. He delineated a difference between a subject and an identity; as a subject, it is about what we do and what we refrain from doing (i.e., how we are, or how we are trying to be). It is consistently emergent, in dialogue with the world, through movement, enskillment, and attentionality, specifically the attentionality that arises in the interdisciplinary space between art and science. The world can create a resistance to that dialogue, which can be frustrating. At the extreme ends of that resistance, the world can be destroyed because we push too hard, or the self can be destroyed because we give up on our ambitions and 'disappear from the world'. Life becomes a balance of existing between these two polarities.

3.3 Part 2: an artist-in-residence

In the next section, Noel Hefele provides a narrative reflection on his experiences as the engaged artist-in-residence. It will provide his voice on the engagement and activities as well as the artifacts as they emerged, unfolded, and developed in his interactions with the young Barbudans.

3.3.1 Reflections on the engaged artist residency

As a location frequently cited as an 'early victim' of the consequences of climate change, they are participants in the culture that is primarily driving this 'world destruction' but not in any significant pollution or CO_2 producing way. Essentially, '[T]he costs and benefits of uneven development are thus distributed unequally: those whose subjugating and over-consumptive stance to "nature" causes the greatest pollution are not the ones who pay its price' (Haritaworn 2015, 211). It can be argued that Barbudan youth may exist in both polarities at the same time.

Before I went to the residency, many folks did not even know the island exists. 'You mean Barbados? Bermuda?' In Barbuda, the youth would ask if my friends could even locate their country on a map. During one interview, one of the youth musicians said he wanted to 'put Barbuda on the map'. We all have the drive to be seen, and art can provide an avenue to satisfy this, to see the world and to be seen in the world (Challe 2015; Hefele 2013, 2015; Russell 2018). Biesta (2017, 16) says that this middle space between 'world-destruction and self-destruction is therefore a thoroughly *worldly* space ... a thoroughly *educational* space that *teaches* you something that is fundamental about human existence, namely *that you are not alone*'.

I, Noel Hefele, primarily paint the local landscapes that I live in. Painting is a way to know that landscape. Questions inevitably arise like 'What Tree is that?' 'Why did it catch my eye?' 'Why is this important?' After looking so intently, I feel a responsibility to know more about the subject of the painting. I also learn more about the landscape from the responses and stories other people share when looking at the work. The story of our relationship with the natural world is written in the language of landscapes. Painting becomes my dowsing rod, where I am drawn to a particular scene, not knowing exactly why. Then through a process of questioning and research, as well as releasing a painting into the world and observing and listening as it finds its way, I uncover natural/cultural information and values that further deepen my connection to that place. The knowledge becomes embodied through the painting practice. I believe the painting also becomes an expression of that place itself.

I didn't know what to expect in Barbuda as I had never been to the Caribbean. I had experience doing community workshops, and I was comfortable speaking about ecological issues but cautious about being the character who parachutes into a community advocating for a 'particular way' of viewing the world. It was a great opportunity to go to a new place and figure out how to collaboratively produce murals around the environment, identity, and climate change while experimenting with informal science and art education during the process.

In Barbuda, it was important to engage in dialogue with the students first and foremost. This helped develop a sense of collaboration and comfort that went against what I sensed was at times a very hierarchical educational relationship with teachers. I wanted to be more of a collaborator with the students. While I am not an anthropologist, I followed Ingold (2013, 2) where '[I]n anthropology ... we go to study with people. And we hope to learn from them'. Ingold also mentions that fieldwork is a process of undergoing an education of attention, and I was attempting to learn how to see the world from their perspective while recognizing that I was to facilitate expression from somewhat of a position of a specialist. I was working against being perceived as a specialist, as I had hoped to work with anyone who expressed interest. 'I'm not an artist' was a common refrain, to which I would always reply, 'me neither'. I was attempting to empower a sense of 'I can do this too', which can be surprisingly difficult with teenagers in general.

3.3.2 *What it means to be Barbudan*

The first mural originated from discussions I had with students at the Secondary school. I went to the Secondary school to do a few landscape painting workshops and attempted to recruit interested folks to work on the mural. After a brief discussion about murals as an art form, along with a few examples, we generated a list of words that might answer the question, 'What does it mean to be Barbudan?'. We filled up the entire chalkboard with generated words, locations, and concepts. It was helpful for me, as an outsider, to get a picture of how students saw their world. I did a series of observational oil painting landscape workshops out on the grounds of the Secondary school. I also shared some images of my own work. I did a PowerPoint presentation highlighting Black and Caribbean artists in order to both show different styles of art and artists that share similar cultural backgrounds and identities. I tried to encourage them with their personal projects. Many of the students expressed a lack of confidence in their artistic ability; however, there were many examples of talent and vision. I worked to encourage interesting elements I saw but also tried to advise more formal and technical elements that may remove friction as they learn about developing an art practice. In some of the oil painting workshops, this involved teaching them about the effects of too much paint too quickly – the colors will get muddy because the oil paint doesn't dry quickly. Some students began to muddy up their pictures and became frustrated. So, we then worked subtractively by grabbing sticks to remove paint or trying various other ways to remove some of the excess paint. One student went from a very tactile engagement with the oil paint during a workshop to help out with several murals; she dreamt excitedly about continuing to make murals herself and leading a small group of students to create a mural for a local commercial establishment. Several students painted to a certain point and then wiped the record clean, self-censoring in either 'I'm too cool for this' or 'I'm not comfortable making artwork' way. Another student gravitated toward a precise style of work, proving to be an essential draftsman for drawing the letters on the final mural. Some students just 'played', having fun and horsing around. One student asked to continue to work on her painting past the end of class, later smiling, saying she took it home and her father said he is 'proud of her'. Not all students engaged deeply, however, a good number did, either with quiet focus or expressive enthusiasm.

After stressing that 'you can do this' and Barbuda itself is a fine theme for art, it followed that the theme of the mural was 'What it means to be Barbudan'. We designed a mural that highlighted the beaches of Barbuda with some additional symbols added to the landscape. The mangroves were included as a habitat for organisms and protection of the shoreline from waves and hurricanes. The frigate birds were included, a colony of birds that Barbuda is well known for. A supply ship was also included as this is critical for providing the island with much-needed food, water, and other resources. And of course, the mural included beautiful beaches, ocean, and sunset. Students finished the mural by painting local organisms that live in the mangroves on the columns surrounding the mural.

They had internalized the knowledge of the organisms and their approximate shapes, easily filling up the columns after quizzing each other on what lives in the mangroves and correcting each other on the shapes of the fish. There were many conversations about the importance of the mangroves for habitat and storm protection. We discussed how the root structure forms a protective barrier that breaks up the kinetic energy of the ocean during storms. We discussed how the roots also bind the soil and protect against erosion. We thought about why the roots might be above the water line to get oxygen since the plant grows in stands of salt water. We discussed how the plant might have some way of dealing with the salt from the water, as not every plant could grow in such conditions. The mangrove is seemingly abundant in Barbuda, but we developed a sense of how special the plant is and how valuable an ecosystem service it provides before expanding into a larger conservation view. We learned that mangrove populations were under threat globally, and that knowing that can make you appreciate the ones you see. We looked at areas along the shoreline and thought about losing one out of every five mangroves. We discussed the felt responsibilities of conservation, preservation, and restoration of Barbuda's mangroves (Figure 3.1).

3.3.3 Elementary school mural

The second mural was at the elementary school. The challenge here was to get the children involved in something they could paint. Some students from the Barbuda Research Complex (BRC) helped organize the effort. The youth were given chalk and instructed to outline themselves on the wall in energetic poses. They did so swimmingly. The older students helped trace those outlines with paint and fill in the figures. Immediately, the young students identified whose silhouette was whose, and frequently would pose in front of the shape in a similar pose. The older students were then instructed

Figure 3.1 The first mural.

to generate a list of words to paint into the mural, make lists of words related to Barbuda and climate change, and then add them in among the characters.

Several of the mottos of the school emerged on that list; 'Be Respectful', 'Be Prepared', 'Be Good', and 'Be Kind'. Curiously, 'Hurricane' or 'Storm' did not show up on that list, but 'Earthquake' did. I found that surprising and almost thought it was unfounded until I remembered the terrible earthquake in Haiti only a few years before and realized that earthquakes are a common occurrence in the Caribbean Basin. The list of words went on to include 'Sun', 'Land', 'Sea', 'Teacher', 'Red Cross', 'Soil', 'Healthy', 'Preserve', 'Fresh Water', 'Earth', 'Drought', 'Earthquake', 'Education', 'Survive', 'Parents', 'Jobs', 'Co-operate', 'Medication', and 'Responsible' (Figure 3.2).

> The more practised we become in walking the paths of observation, according to Gibson, the better we are to notice and to respond fluently to environmental variations and to the parametric invariant that under-write them … That is to say, we undergo what he called an 'education of attention'.
>
> (Ingold 2018, 31)

Figure 3.2 Children in front of mural painted on school building.

We learned a bit through this process of reflecting on the work. In many ways, it was just a beginning, since the successful elements of the effort were not planned for or even immediately recognized, or perhaps not even recognized. I had hopes of pulling more of the individual voices to create murals but had to walk the line of keeping them focused on climate change and maintaining interest and confidence in expressing themselves.

During the second visit, I aimed to foster the capacity of several young people to develop the confidence and ability to paint a mural on their own. Byron's Cafe by the wharf was looking for a mural, so one of the teenagers led a group of younger kids in creating a mural with signage for his wall. They did not finish by the time I left, but later in the pictures I noticed that the mural was complete.

There was a tension that I was not sure how to resolve. Many of the students seemed to position me as the teacher with something to impart, and I was resistant to this. I was aware that I knew next to nothing in this context but not standing on clear enough ground to realize we were on this path *together*. I didn't know the outcome of the efforts, yet self-consciously felt this was a weakness. In retrospect, this not knowing is part of the process and in many ways, I was fully improvising my way through by engaging with the students; this allowed for the music element of the residency to emerge, as we will discuss in the sections to follow. Perceptual tuning through world-facing attention, in other words, 'habit', is an underlying skill of good art and science,

> the principle of habit, on the contrary, rather than starting from ends, produces beginnings. Its creativity is that of 'doing undergoing' of *agencement,* in which beings continually forge themselves and one another in the crucible of social life, their humanity not a foregone conclusion but an ongoing relational achievement.
>
> (Ingold 2018, 33)

3.3.4 *The Sea Will Rise, Barbuda Will Survive*

In 2014, I returned to Barbuda for a second residency in order to work on other murals. My timing for working with the students was off, as they were in the middle of a school test week. I would ask for help with designing the mural; we were aiming to think through the theme of climate change. The students said they would help, but after the testing period. The tests were Caribbean-wide standardized testing, and I perceived that the students were under pressure to perform well. Barbuda was also in the middle of a fairly significant drought during this time, which also added to the collective stress of having a dwindling fresh water supply. However, I was able to have some informal discussions with some of the Secondary school students I met on my prior trip. One discussion, in particular, that happened on the wharf one evening led to the design for the third mural 'The Sea Will Rise, Barbuda Will Survive'.

I was talking with Kevin, a student, about sea level rise in particular. He was 16 at the time. We were thinking through the impacts of sea level rise and projecting out to when Kevin would be 66 years old. Even the slightest sea level rise would allow the ocean into the lagoon, fundamentally disrupting the ecosystem and changing the character of the place where we were sitting. In 50 years, I mentioned that where we were sitting might even be underwater. 'What would you do? What would Barbudans do? What does it mean for Barbuda?' I asked. Kevin shrugged and responded in this extraordinary moment of indifference AND resilience: 'We will survive'. The discussion then veered into questions of 'What would you say to yourself in 50 years? What would you say to your younger self when you are 66?' to which the initial answer was repeated in several forms. The concept of communication across a gulf of time led to the postcard idea, which we agreed was a cool form. The tagline emerged by itself almost immediately. It had the perfect balance of indifference and resilience that Kevin expressed. With the form in place, further discussions with students revolved around 'what are important parts of Barbuda to include in each letter?' We all selected aspects to portray in the landscape: horses, the aquaponics station, Frigatebirds,

Figure 3.3 Sea Will Rise, Barbuda Will Survive mural.

fruit, fish, beaches, and sunsets. It became a postcard to the future, one that we didn't know would be arriving a mere three years later. While the wall on which we painted didn't totally survive Hurricane Irma, the mural did (Figure 3.3).

Climate change is a problem of being in the world, if we teach skills and provide an opportunity for opening up the senses to the world from an empowered stance, rooted in care and connection for the place, we can face the specter of climate change with more resilience.

3.3.5 Tunes of identity and resilience

The music element of the work emerged during a workshop I was giving at the Secondary school. We were all painting out in the landscape around the school. One student, 'Bamma' Jahrocker Russell, was visibly not interested in painting and busied himself with math exercises. I asked him why and he said he was just not interested. I asked him what he was interested in, and he said 'music'. I shared with him that I've also made music on my computer and had some experience creating beats, recording songs, and shooting videos. We decided to try and make a song.

3.3.6 *Life Ain't Easy*

Bamma gathered up two other friends in his crew, 'Vertile' Tommy Joseph and Llemuel 'Rue' John. They called themselves the 'Danger Squad/Vital Sound'. We listened to a bunch of instrumental beat ideas, picked one, and decided to flesh it out. I encouraged them to write from their 'unique' voice; nobody else in the world has the Barbudan perspective, as there can be a tendency in music to mimic common themes that transmit through pop culture. The song they wrote is entitled 'Life Ain't Easy'. It's a very catchy song and very profound. Their lyrics betray their youth with a wisdom beyond their years:

> 'Life Ain't Easy –
> You don't know the things that I'm going through,
> I don't really know what I'm gonna do.
>
> Got my Hoodie on,
> looking like Trevon Martin …'

We took two digital cameras to shoot the video. We would pass the cameras between each other as the song progressed, and each person performed lip sync for the cameras while an mp3 of the song played in the background. We wandered around, shooting in different locations. They did not yet want the attention of their classmates so we chose to be clandestine about creating the video. I took the footage back with me to New York City to edit and post on YouTube.

3.3.7 *Without You*

When I returned in 2014, there was enthusiasm to work on more music videos. BRC-Graduate Center field school students wanted to collaborate, and some other Barbuda kids wanted to as well. But with more hands in the pot, it was a bit more difficult to manage while also mapping out the parallel mural efforts. We held a few listening sessions, trying to find the beginnings of a song. One of the musicians from 2013 had a newborn baby. Another had relocated to Antigua. There were a few new faces in some of the listening sessions, but I could tell the unusual style of some of the music tracks I had as 'starters' were not capturing their attention. To be honest, I felt a bit worried that I would not be able to guide producing the appropriate musical backing for the next project. The youth had grown more sophisticated in their tastes over the past year, and I had not practiced musical production at a pace to keep up. My computer was one year older and somehow much slower. I felt caught in between the expectations of the now one-year older musicians and the expectations of the BRC-Graduate Center field school students. Bamma was very patient and persistent in trying to make it work.

Much of my music follows a hip-hop form, while his interests had diverged to a more Soca or dancehall form. He explained the rhythm patterns of both. I made a few attempts with tracks that he laughed at because I could not quite produce the right vibe. But one working session resulted in the instrumental heard in the song – Bamma explained how to change and manipulate the rhythm while Tiffany, a BRC-Graduate Center field school student, crafted a piano-like melody over the top. Suddenly the track emerged.

I arranged the music on the computer using Ableton Live,[1] trying to foreground other's input as much as possible. This was a great example to see how Bamma could carry a whole song; Vertile and Rue were not available to work together as scheduling issues arose. Tiffany Challe wrote the main melody of the song and sang the hook. Oscar Lemus was the videographer. I helped direct some of the kids that gathered around to enthusiastically jump in and be in the video. One of them borrowed my hat.

This song is a bit of a love song. I asked Bamma if it potentially was about Barbuda, and he laughed but did not dismiss the idea, saying 'Maybe'. I would try to encourage the musicians to make songs about things other than women, money, and fame. These types of songs can be nice, but I was hoping to help develop their unique voices. My main interest resided in using art to tell stories about Barbuda and their unique position in the climate change discussion. I kept stressing that very few people on the planet have their perspectives and that they should write from a place of honesty. This song didn't quite engage on that level, but it was still a catchy tune. At the time, I was worried about that and perhaps pushed on the climate change angle a bit much; there was a point when I could become too didactic. But it is quite possible the third video would not have emerged without these discussions.

3.3.8 *We Do It for Barbuda*

This song was created one afternoon while working with Bamma and Vertile. Vertile took the lead in crafting the instrumental backing, asking me how to do things in the Ableton Live program, and telling me how to arrange the song. I was mostly hands-off on the music production for this instrumental track, other than answering questions and helping them find a particular sound. They wrote lyrics to it for an hour or so, and then we recorded their vocals. They now had a proper microphone that they shared with the larger circle of friends. In 2013, they recorded using a cheap plug-in microphone we found in the research center. Now they had big headphones and a screen for their microphone to guard against popping 'p' sounds when recording, artifacts of improved production, and commitment to craft. We shot the video again while walking around. I took the footage back to New York City and edited it, sending Bamma a few edits along the way.

Again, this song seemed wise beyond their years. And it becomes a different song post-Irma.

> 'Ain't thinking bout my past, I'm thinking bout my future-
> So when my son asks me "Hey Daddy who you do it for?"
> The Music? We do this for Barbuda.
> Hard Work? We do this for Barbuda'

> 'To see my island suffering,
> that's one thing I can't stand
> The love of my country?
> That makes me who I am
> And when I think I can't,
> They say you know you can,
> stand up and be a man,
> I do it for my fans.'

The group was now known as 'Music is the Future MITF' (Music is the Future, n.d. Facebook). When asked what it means, they expressed an amazingly optimistic and resilient view about music and its importance to them. Bamma told me he wants to 'have music put Barbuda on the map'.

In 2014, some of the little kids would have tracks by Bamma on their phones. Many had the lyrics memorized. Bamma told me that they use Bluetooth to transfer from device to device because it is quicker than using the Internet connections they had at the time.

I still communicate with the group via Facebook and Soundcloud. The group is now in Antigua for the most part. Bamma is working on new material, with a real recording studio and need to finance some of that recording time. They all perform at the annual music festival. I've asked him if he'd like to send any newer material over, and he said, 'everything was lost in the storm.' I'm trying to encourage him to make a song about the Irma experience. Bamma has been on the Antigua radio with his new song 'Real Hard'.

This makes me wonder about something I have taken for granted and yet is critical for me as an artist; I began to think about the Barbuda-wide impact on electronics and data storage. How much data was lost in the storm? Younger people exist heavily on their digital devices. What is the impact of losing all of that overnight? Losing data is a catastrophe for most of us. If your artwork is only stored digitally, how to ensure its resilience? In addition to opportunities to work expansively and open-ended, it was clear and sobering that some practical considerations, like ensuring the work survives, are also valuable in this context.

3.3.9 *Post-Irma*

Watching from New York City, it was very surreal to see the images of Barbuda from space right before and after Hurricane Irma. The island turned from green to brown, as apparently all of the vegetation was blown away. It's an understatement to say the island was devastated by the storm. Images being published on news sites appeared post-apocalyptic. The storm destroyed the musicians' files, and it threatened the structures the murals were on themselves. The intended subject of the artistic efforts spoke loudly and threatened the existence of the work itself. In many ways, it was luck that the wall with *The Sea Will Rise, Barbuda Will Survive* mural was not knocked down. It was also lucky that we stumbled upon a mural that spoke so directly to the new context it was in, at least from my distant point of view in Brooklyn, New York.

After Irma, *The Sea Will Rise, Barbuda Will Survive* mural appeared on several Barbudan social media accounts along with countless images of the island's devastated structures. The context around the mural had changed for me. I no longer thought that perhaps the mural was too didactic or disconnected from the lived everyday lives of the youth that helped create it. It was inextricably a part of the fabric of the landscape and *of living in* Barbuda. Mohammid Walbrook wrote on October 8, 2017, regarding the mural, 'this should be our new motto going forward with the clean-up and rebuilding od [sic] BARBUDA' (Walbrook 2017). We had anticipated that the world would speak to Barbuda in this fashion at some point, but the form of the mural was ontologically open to the world it grew into, hence provoking the world to grow into it.

The context around the other two murals did not seem to change with the storm. I think the mural of the children at the school may have gotten painted over. *What It Means to Be Barbudan* seemed to primarily deal with symbols answering the question posed by the title. Symbols are fine, but they are distilled abstractions, solid and predefined containers of meaning, not fully suited for corresponding with the uncertain, complex entanglements of a shifting, climate-changing world. Hurricane Irma seems to seal off *What It Means to Be Barbudan* from the present; the mural becomes an artifact of the time before the storm unlike *The Sea Will Rise, Barbuda Will Survive* mural, which takes on new meaning and importance after the storm. The mangroves initially depicted and conceptualized as protectors now seemed outmatched by the ferocity of the hurricane.

The outline mural at the school primarily worked along the identification lines. Students could see themselves and others in the image. They would embody the outlines of others while also getting to embody the process of making the mural themselves. It felt like more of a participatory exercise in this regard, not without its own value, but perhaps the method and content could be improved by iterating on the idea of outlining each other.

The success of *The Sea Will Rise, Barbuda Will Survive* mural as artwork was luck, unforeseen, and open-ended. Yet, recognizing how it was successful in this way that the others were not brings up interesting questions on how to integrate this essence when prompting students to create something open to and in the world. But, how do you aim for the unforeseen? The most successful artworks grow and change, participating with the world and its inhabitants.

3.4 Part 3: connecting practice with theory

If we return to the third definition of resilience offered up by the England Arts Council (Woodley et al. 2018), **Resilience as 'bouncing forward'** can be thought of as applying to either of the first two murals. The resiliency in **what it means to be Barbudan mural** lies somewhere in between absorbing and bouncing forward from the disturbances or shocks of a storm surge. The mural with the kid outlines at the school also engages with these definitions of resilience in some fashion, mainly through the young children being able to adapt to the challenges of the world they are growing up in. Both murals do not go beyond conceptualizing resilience in a reactionary mindset. The threats are from the outside. The mangroves offer protection from storm surges and erosion of the shoreline: an ecological resource worthy of appreciation and preservation. The mural at the school approaches the psychological tenacity and expectations of the youth. In both instances, they are conceptualizing something that *already exists*. This is important in terms of solidifying identity but is less helpful in developing agency and resilience in the face of an unknown shifting world. Reactionary responses don't include the sense of agency we feel is important.

Resilience as 'positive adaptability' in anticipation of, or in response to, shocks is demonstrated in *The Sea Will Rise, Barbuda Will Survive* mural. By stating the inevitability of the sea rising, it expresses anticipation of the shock, and by anticipating a shock that is unknown you are able to create the imaginary space to formulate some kind of response; it is a space of possibilities. Stating with clarity that Barbuda will survive represents positive adaptability, and contained within the word Barbuda are symbolic landscapes that represent values that will aid that adaptability. Creating this mural anticipated the conditions of something like Irma at some point, like a postcard across time. Irma may have changed life in Barbuda forever. The island is still Barbuda, but it is not the same. Life is changing, growing, and altering at an ever-quickening pace. The best art is open to the world and entangled with it, changing and growing along with us over time.

3.4.1 Re-making their world

We had no expectations of exactly how these residencies would turn out. There was an opportunity to approach a vulnerable community with an informal art and science pedagogical intention to engage with climate change. The process was a learning experience in its own right, but the final video and final mural effort seemed to reach a level of some kind of worthwhile impact as it speaks to the innate resilience of the Barbudan youth. While there are many similarities in teaching global youth to think about climate change, the youth of Barbuda draw certain responses about what we can do into stark relief. It makes no sense to say 'well change a light bulb' or 'take shorter showers'; their way of life is already one of significantly low impact, yet they face significant impact from climate change. We cannot simply transmit facts about climate change to students, because as Timothy Morton (2018, xxii) writes,

> [F]acts go out of date all the time, especially ecological facts, and especially, out of those, global warming facts, which are notoriously multidimensional and scaled to all kinds of temporalities and all kinds of scenarios. Dumping information on ourselves every day or every week can be really confusing and arduous.

Students should learn to correspond with the world through a process of making. Through making art, students begin to establish their place in the world and make statements about what they want to project for their futures. In this case, the Barbudan youth generated art that projects resilience, persistence, and perseverance.

Climate change is an increasingly dire specter clouding over all of our futures, but somehow we as humans will survive. Figuring that out resides somewhere in an open-ended imaginary space, one that anticipates a changing and shifting, yet unforeseen world. This new space is world facing, committed to dialogue or correspondence with that world. We believe that an integrated art and science pedagogical space can provide ways to speak to that world and listen when it speaks back, paying attention to and reflecting on what it has to tell us. Science and art are well suited to approach the uncertain future of climate change because they are both practices comfortable in not knowing.

Note

1 https://www.ableton.com/en/.

References

Adams, J.D. and Gupta, P., 2017. Informal science institutions and learning to teach: An examination of identity, agency, and affordances. *Journal of Research in Science Teaching*, 54(1), pp.121–138.

Adams, J.D. and McCullough, S. in press. Inquiry and learning in informal settings. In C. A. Chinn & R. G. Duncan (Eds.), *International Handbook on Learning and Inquiry*. New York: Routledge.

Adams, J.D., Fortwangler, C. and Gibney-Sewer, H. 2017, 19 October. Green islands for all?: Avoiding climate gentrification in the Caribbean [Blog post]. Retrieved from https://ethnobiology.org/forage/blog/green-islands-all-avoiding-climate-gentrification-caribbean.

Biesta, Gert., 2017. What if? Art education beyond expression and creativity. In *Art, Artists and Pedagogy*. Routledge, pp.29–38.

Biesta, G., Priestley, M. and Robinson, S., 2017. Talking about education: Exploring the significance of teachers' talk for teacher agency. *Journal of Curriculum Studies*, 49(1), pp.38–54.

Biesta, G. and Tedder, M., 2006. *How Is Agency Possible? Towards an Ecological Understanding of Agency-as-Achievement. Learning Lives: Learning, Identity, and Agency in the Life Course.* Working Paper Five, Exeter: Teaching and Learning Research Programme. Publisher: The Learning Lives project, Exeter http://hdl.handle.net/10993/13718.

Challe, T., 2015. Without you – Barbuda music video Bamma ft. Tiffany Challe. *Youtube*, lyrics, music production and by Bamma and Tiffany Challe; music production by Noel Hefele, video production by Oscar Lemus, May 15, 2015, youtu.be/Ev_xAkT-jTY.

de Sousa Santos, B., 2007. Beyond abyssal thinking: From global lines to ecologies of knowledges. *Review (Fernand Braudel Center)*, pp.45–89.

Douglas, A., 2012. Altering a fixed identity: Thinking through improvisation. *Critical Studies in Improvisation*, 8(2) pp. 1–12.

Freire, Paulo., 1998. *Pedagogy of Freedom: Ethics, Democracy, And Civic Courage.* Lanham: Rowman & Littlefield Publishers.

Haritaworn, J., 2015. Decolonizing the non/human. *GLQ: A Journal of Lesbian and Gay Studies*, 21(2), pp.210–213.

Hefele, N., 2013. Life ain't easy- danger squad / vital sound (vertile, rue and Bamma). *Youtube*, lyrics, music arrangement and video shooting by Vertile, Rue and Bamma; guidance and video editing by Noel Hefele, April 26, 2013, youtu .be/W6osPgvKr8w.

Hefele, N., 2015. We Do It for Barbuda – MUSIC VIDEO – by Music is the Future - MITF. *Youtube*, lyrics, music production, video direction by Bamma and Vertile; production guidance and video editing by Noel Hefele, April 30, 2015, youtu.be /82JO_S4iWBs.

Ingold, T., 2011. *Being Alive: Essays on Movement, Knowledge and Description.* Taylor & Francis.

Ingold, T., 2013. *Making: Anthropology, Archaeology, Art and Architecture.* UK: Routledge.

Ingold, T., 2018. *Anthropology and/as Education.* UK: Routledge.

Levy, B.S. and Patz, J.A., 2015. Climate change, human rights, and social justice. *Annals of Global Health*, 81(3), pp.310–322.

Mohammid Walbrook post. *Facebook*, Oct 8, 2017, https://www.facebook.com/ mohammid.walbrook/posts/10156651448033272. Accessed January 27, 2021.

Morton, T., 2018. Introduction: Not another information dump. In *Being Ecological.* Cambridge Massachusetts: MIT Press, pp.xiii–xlii.

Music is the Future. n.d. *Facebook* [Fanpage]. https://www.facebook.com/MusicIs TheFutureMITF/.

Russell, Jahrocker Bamma. *Soundcloud*, Nov 6, 2018, pg 74. https://soundcloud .com/bamma-jahrocker-russell.

Woodley, S., Towell, P., Turpin, R., Thelwall, S. and Schneider, P., 2018. What Is Resilience Anyway? A Review. Commissioned by Arts Council England. https://www.artscouncil.org.uk/sites/default/files/download-file/What%20Is %20Resilience%20Anyway.pdf.

4 Fallow deer

The unprotected biocultural heritage of Barbuda

Naomi Sykes

4.1 Introduction

In 1963, Zeuner set out his five stages of domestication, in which he suggested the last stage to be the 'persecution/extermination of wild ancestors'. His statement was made just two years before the International Union for the Conservation of Nature (IUCN) founded its Red List of Threatened Species. This was set up to halt the extinctions that had already befallen the wild ancestors of modern domestic cattle, horses, dromedary camels, and Guinea pigs (Dobney and Larson 2006) and continue to threaten others (e.g., the European rabbit, wild cat, gray wolf). In the 21st century, the need for monitoring and protecting wild animal populations is becoming all the more pressing as the world is under growing pressures from increasing human population and the associated intensification of food production, urbanization, globalization, and environmental degradation. For these reasons, the preservation of biodiversity features prominently in many of the Sustainable Development Goals (SDGs) set out in the United Nation's *2030 Agenda for Sustainable Development*.[1]

A wide variety of strategies are being put in place to conserve wildlife, with 'conservation translocation' becoming an increasingly popular method in restoration ecology, whereby animals from one population are relocated to replenish those that are under threat of extinction or have become extinct (Batson et al. 2015). Paradoxically, translocation is also a criterion for denying species entry onto the IUCN Red List, with human-assisted movement deemed to represent a level of anthropogenic management that negates an animal's classification as 'wild'. Beyond being exclusionary in terms of conservation policy, many conservation scientists argue that species translocations, or 'introduced animals', are one of the greatest threats to global biodiversity, responsible for outcompeting or hybridizing with endemic populations (e.g., Firn et al. 2015; Jackson 2015; Wyatt et al. 2008).

In many cases, judgments concerning the negative impact of introduced species are valid. For instance, colonial introductions to Australia had a devastating effect on native wildlife (e.g., Shine 2010; Woinarski et al. 2011; 2015) and similar scenarios can be found around the world (e.g., Nellis and

DOI: 10.4324/9781003347996-5

Everard 1983; Pimm et al. 2006). However, these high-profile examples have led to the emergence of conservation policies that, although purporting to be purely scientific (the subtext being that decisions are therefore a-cultural), are not only stereotypical but are also laden with value judgments. The labels 'wild' and 'native' are equated with pristine and natural and are therefore perceived as inherently good. In stark opposition are the 'feral' and 'introduced' or, worse, the 'alien' and 'invasive', all presented in negative terms. In much the same way as anthropologist, Mary Douglas (2003) defined perceptions of dirt and pollution, and introduced fauna are widely vilified as 'animals out of place'.

It has not gone unnoticed that the language employed in biodiversity and wildlife management discourse is startlingly similar to that found in right-wing nationalistic rhetoric, both seemingly preoccupied with native versus alien status, and emphasizing the need for 'genetic purity' of populations. This similarity has generated heated debate and division within conservation biology, with those who wish to eradicate introduced species being labeled as xenophobic, whilst those opposing eradication are seen as promoting biological homogenization (e.g., Coates 2006; Frawley and McCalman 2014; Preston 2009; Skandrani et al. 2014; Shackleford et al. 2013; Warren 2007). However, all of this contention is frequently based on modern datasets, with little deep-time empirical evidence concerning the actual native/alien status of the species under consideration. Archaeological studies have the potential to remedy this, and they are increasingly showing how little we know about biodiversity.

For instance, in Britain, the brown hare (*Lepus europaeus*) is widely considered both wild and native and as such is the subject of a variety of conservation efforts (e.g., Game and Wildlife Conservation Trust, Hare Preservation Trust). Yet archaeological studies have demonstrated that the brown hare is neither native nor truly wild; it was introduced to the island around 2,500 years ago, was farmed for approximately 500 years during the Roman period, and only subsequently escaped to form feral populations (Lauritsen et al. 2018). Similarly, the Przewalski horse (*Equus ferus przewalskii*), as its Latin name implies, has long been considered to be the wild progenitor of modern-day domestic horses. It is listed as Endangered on the IUCN Red List with a detailed conservation plan in place that includes captive breeding programs at many international zoos (King et al. 2015). Recent genetic studies comparing modern and ancient DNA of global horse populations have revealed that, far from being a pristine wild population, the Przewalski horse is the feral ancestor of an ancient domesticate (Gaunitz et al. 2018).

These two examples reveal that modern conservation policy is frequently founded on very superficial, often highly fallible, knowledge. Moreover, it raises the question of whether the feral status of both the brown hare and Przewalski horse renders their conservation less urgent or important. The answer must surely be no, as both are important keystone species from a cultural perspective. In which case, it is perhaps time for us to rethink our values and consider if, rather than judging a species on its perceived native/wild/feral/introduced status, it be considered on the basis of its ecological function (as argued by Davis et al. 2011; Schlaepfer et al. 2011) but also cultural significance (e.g., Fortwangler 2009; 2013).

The European Fallow deer (*Dama dama dama*) serves as an excellent focal point for this kind of discussion. The species has a long, interconnected history with humans, and its status is complex; across the world today, it is simultaneously considered to be locally extinct, endangered, introduced, invasive, and a cultural icon. The latter is true for Barbuda and Antigua, where the fallow deer is the national animal, so synonymous with the islands it is represented on their coat of arms (Figure 4.1, McComas 2013, 126). However, as an introduced species that is considered neither domestic nor wild, it falls outside any formal protection. As such, the population is fragile, with the potential of being eradicated due to overhunting or as a result of natural disasters.

4.2 Fallow deer of Barbuda: their origins and biocultural significance

The fallow deer is the most widely distributed cervid on the planet, having been transported around the world by humans over the last 10,000 years. The species is native to the eastern Mediterranean but was taken to many Aegean and Mediterranean islands during the Neolithic and Bronze Ages. Further

Figure 4.1 Antigua and Barbuda's national coat of arms (Consulate General of Antigua and Barbuda).

waves of introduction during the Roman and medieval periods took fallow deer to mainland and northern Europe (Sykes et al. 2006; Madgwick et al. 2013; Valenzuela et al. 2016). Their subsequent diffusion around the world was the result of European colonialism, which saw fallow deer established in South Africa, Australia, New Zealand, USA, Canada, Argentina, Chile, Peru, and Uruguay, as well as islands in the Fijian group, and the Lesser Antilles (Chapman and Chapman 1980). In many of these countries today, the fallow deer are managed and even farmed. Its status as a semi-domesticate is reflected by the variety of pelage types displayed by the species – common, menil, melanistic (black), leucistic (white), and long-haired (Masseti 1996). Coat color variation is recognized as a trait of domestication (Linderholm and Larson 2013), and the majority of translocated deer populations, including those on Barbuda, exhibit different pelage colors (Figure 4.2).

In keeping with Zeuner's (1963) suggestion that the final stage of domestication is the wild progenitor's extinction, there is evidence to suggest that fallow deer have become locally extinct across the Balkans, and the last remaining Anatolian wild/native population of fallow deer is under threat. It is estimated that a maximum of 120 individuals (and potentially just 96) survive within the UNESCO World Heritage Centre of Gulluk Dagi-Termessos National Park, Turkey (Ünala and Çulhacıa 2018). The IUCN Red List suggests that conservation measures are needed for the Termessos population and recommends that protection be extended to herds that inhabit the island of Rhodes (Masseti and Mertzanidou 2008).

The fallow deer of Rhodes are thought to descend from animals introduced during the Neolithic period (Masseti et al. 2006) and have become cultural icons, their images depicted on artifacts and street signs throughout the island. More notably, a statue of a fallow deer buck has replaced the ancient wonder of the world, the Colossus of Rhodes, which welcomes visitors arriving at the island's harbor. The species is protected by the Greek

Figure 4.2 Coat color variant seen in the Barbuda deer population. Photo source: (Perdikaris et al. 2018)

law, with a government-funded book, *The Island of Deer*, written about their biocultural significance (Masseti 2003).

In many ways, the fallow deer of Barbuda are similar to those of Rhodes in that they are not native but have, nevertheless, become an important element of the island's economy and cultural identity (Perdikaris et al. 2018; Baker et al. 2015). Unlike the Rhodes population, however, the Barbuda deer have little to no protection largely because they were introduced in recent centuries and are therefore considered an 'alien' species.

Based on historical and genetic evidence, Perdikaris et al. (2018) have argued that fallow deer were imported to Barbuda from England in the late 17th or early 18th century, most probably by the Codrington family, who were granted the first lease on Barbuda in 1684 (Murray 2001, 66). The Codringtons brought not only fallow deer to Barbuda but also hundreds of West African slaves who were worked to supply provisions for the Codrington's sugar plantations on Antigua (Lowenthal and Clarke 1977). The history of Barbuda's fallow deer and human populations are critically entangled. Both the humans and deer descend from individuals that were transported and exploited under colonial powers. More significantly, fallow deer became an important vehicle through which the Barbudan slaves subverted authority; they poached the Codrington's beloved animals. Documentary evidence records repeated complaints about Barbudan's poaching forays, with some of the most intensive periods of poaching occurring during periods of slave rebellions and uprisings between 1761 and 1790 (Murray 2001, 271). Contrary to English traditions, whereby hunting rights and land ownership were restricted to the elite, Barbudan slaves had a very different attitude, perceiving both the deer and land as no-one's and therefore everyone's property. This sense of common possession prompted one overseer to write to the Codrington family out of frustration 'they acknowledge no Master, and believe the Island belongs to themselves' (R. Jarritt 1823, quoted in Lowenthal and Clarke 1977, 524). Just over a decade later, in 1834, the Barbudan slaves were emancipated and both deer and people have coexisted on the island together since.

The Barbudan heritage embodied by the fallow deer is reflected by the fact that, even today, men still say they are going out to 'poach' rather than to 'hunt' (McComas 2013, 126). Furthermore, fallow deer play an important role in the Barbudans 'Living from the Land' celebrations, which involve hunting, feasting, and socializing (Perdikaris et al. 2013). However, the sustainability of Barbuda's fallow deer herds, as well as the traditional cultural attitudes and practices towards land ownership (as codified Barbuda Land Act of 2007), is under threat following Hurricane Irma (Gould and Lewis 2018).

4.3 Fallow deer management in the aftermath of Irma

Even prior to the hurricane, Perdikaris et al. (2018) warned that Barbuda's fallow deer population was vulnerable due to a lack of management and protection against threats of overhunting. In general, island populations are at higher risk of extinction than those on the mainland, with 71% of modern era global extinctions having occurred on islands (MacPhee and Flemming 1999). Introduced animals are not immune to these forces; however, our deep-time study of fallow deer has shown that the majority of populations introduced to islands in antiquity subsequently went extinct such as those introduced to Neolithic Cyprus and Crete, Roman Mallorca, Sicily, and Britain (see Valenzuela et al. 2016).

Climate change is compounding pre-existing problems for the Barbudan deer. Sea level rise is causing marine infiltration of freshwater sources and extreme weather, such as Hurricane Irma, denudes the vegetation cover upon which the fallow deer rely for shelter and food. Reports from Barbuda suggest that young deer were found wounded and disoriented after the hurricane, with some being sent for veterinary treatment to Antigua (Perdikaris pers. comm.). However, there has been no move to establish sanctuaries for the deer or check the status of their populations.

The lack of attention being afforded Barbuda's fallow deer can be linked to the fact that, within much conservation literature, they are labeled as 'alien invasives', listed among a number of species deemed responsible for the demise of native fauna through competition, disease transmission, and habitat alteration (e.g., Daltry 2007; Borroto-Páez and Woods 2012; Lindsay 2014; Smith et al. 2014; van der Burg et al. 2012). Nowhere, however, is there any explanation about how the label of invasive has been arrived at in the case of fallow deer since there is no evidence to suggest that the *dama* populations specifically are responsible for any damage to local biodiversity. This is by contrast to feral goats and to a lesser extent donkeys which do appear to be responsible for some overgrazing, and domestic/feral cats and dogs are also clearly problematic (Borroto-Páez and Woods 2012). It seems illogical, then, that in the aftermath of Hurricane Irma, the limited animal aid that was deployed focused almost exclusively on the animals that have been most accused of damaging biodiversity. For instance, the Donkey Sanctuary sent teams to ensure the welfare of Barbuda's feral donkey population, whereas the World Animal Protection and the Humane Society went to the rescue of domestic livestock, dogs, and cats. No assistance was sent for the fallow deer because, as a translocated animal, they fall outside the remit of IUCN concerns. They remain, however, the national animal of Barbuda and Antigua, a cultural heritage that is entirely unprotected.

It would seem that Western NGOs are more concerned with animals that are aligned to their own cultural values (pets and livestock) without giving consideration to what is important to the islanders themselves. There is growing criticism of, and reaction against, animal conservation as a form of neocolonialism and neoliberalism (Büscher et al. 2012). For instance, Garland (2008) highlighted how African elephant conservation resulted in power imbalances between conservationists and local groups. Similarly, Hayward et al. (2018) have argued that translocations of African Rhinos to Australia for a breeding program is simply the continued exploitation of Africa's resources by colonial powers. There are an increasing number of initiatives to involve local African groups as well as First Nation and Aboriginal communities in the conservation of their own wildlife (Bhattacharyya and Slocombe 2017; Bach et al. 2019). These moves are highly welcome, but they are also intriguing from a human–animal studies perspective.

It is widely accepted that the way in which people perceive and treat animals is a reflection of human–human relationships (e.g., Mullin 1999) and the situation on Barbuda is an eloquent example of this. In this article, I have suggested that, in animal conservation, native status is everything, with translocation rendering a species unworthy of protection. In human terms, the same would also appear to be true. UNESCO has recently launched a series of statements and documents pertaining to native/Indigenous cultural heritage (for instance, their 2006 *United Nations Declaration on the Rights of Indigenous Peoples* and 2017 *Policy on Engaging with Indigenous Peoples*). However, UNESCO has no declarations concerning the rights of translocated people, such as the Barbudans, who were first brought against their will to the island as slaves and are now evacuated from their homes due to the impact of a changing climate for which they bear no responsibility. Unlike First Nation and Aboriginal groups, the Barbudans are not being asked their opinions concerning the preservation of their heritage, which in this case includes both fallow deer and cultural attitudes towards land ownership. It has been shown that Hurricane Irma has been used as an opportunity to override the 2007 Land Act (Boger and Perdikaris 2019; Gould and Lewis 2018; Perdikaris et al. 2021a, 2021b). Nor are they being consulted by conservationists about which species they would like to protect. In both human and animal terms, translocation, especially if it happened within historic memory, results in loss of power and rights.

A decade ago, Fortwangler (2009) made the case that conservation efforts to remove 'invasive' donkeys from the Virgin Islands were perceived by the African–Caribbean residents of St. John as a move to erase their collective history from the island and its landscape. More recently, Keehner et al. (2016) made similar points concerning the non-native, white-tailed deer on the island of St. Kitts; the species, although introduced in the late 19th/early 20th century, has social and cultural values to the people of the island. Indeed in 1987, white-tailed deer gained legal protection on St. Kitts and Nevis and, while this legislation has proven difficult to implement (Keehner et al. 2016), it is encouraging that legal structures exist. No such protection is in place for the fallow deer of Barbuda, which as the national animal has arguably even greater social and cultural values to the islanders. Without formal legislation, there is a genuine risk that, as with other island populations throughout time and space, the Barbudan fallow deer could go extinct. This would be a devastating loss of cultural heritage, of economic potential, and even of biodiversity, since the fallow deer now add to, rather than detract from, the islands ecosystem. In order to ensure the species' protection, more work needs to be done to evaluate the island's fallow deer population and raise its profile on the international stage. I hope that this chapter represents a small first step towards achieving these goals.

4.4 Conclusion

The preservation of biodiversity is a global challenge that requires a global response. This should not be at the expense of local concerns, which may not map neatly onto the international policy that privileges 'native' status, be it in relation to plants, animals, or people.

Archaeological studies are increasingly showing that few species, even those perceived to be 'native' or 'wild', have been unimpacted by humans. The histories of humans and animals are and have always been entangled. It is important for modern conservation policy to recognize that, in effect, we are tackling legacies of human impact on biodiversity that span millennia.

In the case of translocated animals, there is a need to consider them within the context of their new environments. Rather than instantly equating introduced with 'bad', there is scope for considering some species as living cultural monuments that should be valued rather than vilified. Ancient migrations of people, ideas, and animals are widely celebrated and incorporated into expressions of cultural identity and cultural heritage. However, the more recent the migrations, the more negative the attitudes toward them, with perceptions often translating into societal attitudes and policymaking. Fallow deer exemplify this mindset perfectly. The fallow deer of Rhodes are presumed, but not yet proven, to have been introduced to the island during the Neolithic, and are viewed as a cultural asset protected by Greek law and featured on the IUCN Red List. By contrast, the Barbudan population is labeled as 'invasive' within conservation literature. Their only crime appears to be that they are an introduced species that happened too recently to have acquired a patina of age-based authenticity.

Time depth, therefore, also has a value judgment attached to it (old = good, recent = bad), and this is applied to animals, people, and cultural traditions. The cultural traditions of the Barbudans – their land rights and living from the land festivals – are deemed too recent to be worthy of protection. Gould and Lewis (2018) have highlighted how Hurricane Irma is being used as an opportunity for 'green gentrification', whereby the global elites who are most responsible for climate change use disaster relief as a mechanism for their own capitalist developments. In so doing, the power, concerns, and traditions of the local population are being eroded, deemed irrelevant, and ultimately eradicated. In a similar way, modern conservation ideologies that privilege the native and vilify the introduced, which are developed by global elites rather than developed in the country, may cause fallow deer populations to be quietly denuded on Barbuda.

Our research on global fallow deer populations has highlighted that extirpation events have happened repeatedly in the past and will likely happen again in the absence of sustainable management practices. They serve as a useful deep-time warning, raising the need for a management plan for Barbuda's fallow deer, the island's unprotected cultural heritage.

Acknowledgments

We are grateful to the editors and two anonymous reviewers whose comments improved the manuscript. This work was supported by the Arts and Humanities Research Council [grant numbers AH/I026456/1].

Note

1 https://sdgs.un.org/2030agenda.

References

Antigua and Barbuda Coat of Arms. Consulate General of Antigua and Barbuda. https://www.antiguabarbudaconsulate.com/antigua-and-barbuda

Bach, T.M., Kull, C.A. and Rangan, H., 2019. From killing lists to healthy country: Aboriginal approaches to weed control in the Kimberley, Western Australia. *Journal of environmental management*, 229, pp. 182–192.

Baker, K., Grouard, S., Perdikaris, S. and Sykes, N., 2015. From icon of empire to national emblem: The fallow deer of Barbuda. In *25th Congress of the International Association of Caribbean Archaeology*. San Juan Puerto Rico.

Batson, W.G., Gordon, I.J., Fletcher, D.B. and Manning, A.D., 2015. Translocation tactics: A framework to support the IUCN Guidelines for wildlife translocations and improve the quality of applied methods. *Journal of Applied Ecology*, 52(6), pp. 1598–1607.

Bhattacharyya, J. and Slocombe, S., 2017. Animal agency: Wildlife management from a kincentric perspective. *Ecosphere*, 8(10), p. e01978.

Boger, R. and Perdikaris, S., 2019. After Irma, disaster capitalism threatens cultural heritage in Barbuda, North American Caribbean and Latin America (NACLA), Feb. 11, 2019. Available at https://nacla.org/author/Rebecca%20Boger%20and%20Sophia%20Perdikaris.

Borroto-Páez, R. and Woods, C.A., 2012. Status and impact of introduced mammals in the West Indies. In C. A. Woods, F. E. Sergile (Eds.) *Terrestrial mammals of the West Indies: Contributions*. FL: Wocahoota Press and Florida Museum of Natural History, pp. 241–258.

Büscher, B., Sullivan, S., Neves, K., Igoe, J. and Brockington, D., 2012. Towards a synthesized critique of neoliberal biodiversity conservation. *Capitalism Nature Socialism*, 23(2), pp. 4–30.

Chapman, N.G. and Chapman, D.I., 1980. The distribution of fallow deer: A worldwide review. *Mammal Review*, 10(2–3), pp. 61–138.

Coates, P., 2006. *Strangers on the land: American perceptions of immigrant and invasive species*. Berkeley, CA: University of California Press.

Daltry, J.C., 2007. An introduction to the herpetofauna of Antigua, Barbuda and Redonda, with some conservation recommendations. *Applied Herpetology*, 4(2), pp. 97–130.

Davis, M.A., Chew, M.K., Hobbs, R.J., Lugo, A.E., Ewel, J.J., Vermeij, G.J., Brown, J.H., Rosenzweig, M.L., Gardener, M.R., Carroll, S.P. and Thompson, K., 2011. Don't judge species on their origins. *Nature*, 474(7350), p. 153.

Dobney, K. and Larson, G., 2006. Genetics and animal domestication: New windows on an elusive process. *Journal of Zoology*, 269(2), pp. 261–271.

Douglas, M., 2003. *Purity and danger: An analysis of concepts of pollution and taboo*. Oxfordshire, UK: Routledge.

Firn, J., Maggini, R., Chadès, I., Nicol, S., Walters, B., Reeson, A., Martin, T.G., Possingham, H.P., Pichancourt, J.B., Ponce-Reyes, R. and Carwardine, J., 2015. Priority threat management of invasive animals to protect biodiversity under climate change. *Global change biology*, 21(11), pp. 3917–3930.

Fortwangler, C., 2009. A place for the donkey: Natives and aliens in the US Virgin Islands. *Landscape Research*, 34(2), pp. 205–222.

Fortwangler, C., 2013. Untangling introduced and invasive animals. *Environment and Society*, 4(1), pp. 41–59.

Frawley, J. and McCalman, I. eds., 2014. *Rethinking invasion ecologies from the environmental humanities*. Oxfordshire, UK: Routledge.

Garland, E., 2008. The elephant in the room: Confronting the colonial character of wildlife conservation in Africa. *African Studies Review*, 51(3), pp. 51–74.

Gaunitz, C., Fages, A., Hanghøj, K., Albrechtsen, A., Khan, N., Schubert, M., Seguin-Orlando, A., Owens, I.J., Felkel, S., Bignon-Lau, O. and de Barros Damgaard, P., 2018. Ancient genomes revisit the ancestry of domestic and Przewalski's horses. *Science*, 360(6384), pp. 111–114.

Gould, K.A. and Lewis, T.L., 2018. Green gentrification and disaster capitalism in Barbuda: Barbuda has long exemplified an alternative to mainstream tourist development in the Caribbean. After Irma and Maria, that could change. *NACLA Report on the Americas*, 50(2), pp. 148–153.

Hayward, M.W., Ripple, W.J., Kerley, G.I., Landman, M., Plotz, R.D. and Garnett, S.T., 2018. Neocolonial conservation: Is moving rhinos to Australia conservation or intellectual property loss. *Conservation Letters*, 11(1), p. e12354.

Jackson, M.C., 2015. Interactions among multiple invasive animals. *Ecology*, 96(8), pp. 2035–2041.

Keehner, J.R., Cruz-Martinez, L. and Knobel, D., 2016. Conservation value, history and legal status of non-native white-tailed deer (Odocoileus virginianus) on the Caribbean island of St. Kitts. *Tropical Conservation Science*, 9(2), pp. 758–775.

King, S.R.B., Boyd, L., Zimmermann, W. and Kendall, B.E., 2015. *Equus ferus ssp. przewalskii* (errata version published in 2016). *The IUCN Red List of Threatened Species* 2015: e.T7961A97205530. http://doi.org/10.2305/IUCN.UK.2015-2.RLTS.T7961A45172099.en

Lauritsen, M., Allen, R., Alves, J.M., Ameen, C., Fowler, T., Irving-Pease, E., Larson, G., Murphy, L.J., Outram, A.K., Pilgrim, E. and Shaw, P.A., 2018. Celebrating Easter, Christmas and their associated alien fauna. *World Archaeology*, 50(2), pp. 285–299.

Linderholm, A. and Larson, G., 2013, June. The role of humans in facilitating and sustaining coat colour variation in domestic animals. In *Seminars in cell & developmental biology*(Vol. 24, No. 6–7). Academic Press, pp. 587–593.

Lindsay, K.C., 2014. The ferns of Antigua and Barbuda: A case of resurgence and resilience. *Pteridologist*, 6(1), pp. 14–18.

Lowenthal, D. and Clarke, C.G., 1977. Slave-breeding in Barbuda: The past of a negro myth. *Annals of the New York Academy of Sciences*, 292(1), pp. 510–535.

MacPhee, R.D.E. and Flemming, C., 1999. Requiem Aeternam. In R. D. E. MacPhee (ed.) *Extinctions in near time*. Boston, MA: Springer, pp. 333–371.

Madgwick, R., Sykes, N., Miller, H., Symmons, R., Morris, J. and Lamb, A., 2013. Fallow deer (Dama dama dama) management in Roman South-East Britain. *Archaeological and Anthropological Sciences*, 5(2), pp. 111–122.

Masseti, M., 1996. The postglacial diffusion of the genus *Dama* Frisch, 1775, in the Mediterranean region. *Supplemento alle Ricerche di Biologia della Selvaggina*, 25, pp. 7–29.

Masseti, M., 2003. *Island of Deer : Natural History of the Fallow Deer of Rhodes and of the Vertebrates of the Dodecanese*. Rhodes, UK: City of Rhodes Environment Organization.

Masseti, M., Cavallaro, A., Pecchioli, E. and Vernesi, C., 2006. Artificial occurrence of the fallow deer, Dama dama dama (L., 1758), on the island of Rhodes (Greece): Insight from mtDNA analysis. *Human Evolution*, 21(2), pp. 167–175.

Masseti, M. and Mertzanidou, D., 2008. *Dama dama. The IUCN Red List of Threatened Species* 2008: e.T42188A10656554. http://doi.org/10.2305/IUCN.UK.2008.RLTS.T42188A10656554.en. Downloaded on 17 February 2019.

McComas, M.J., 2013. *Food and drink guide 2013*. St. Johns: Leeward Consultants and Associates Ltd, Woods Centre.

Mullin, M.H., 1999. Mirrors and windows: Sociocultural studies of human-animal relationships. *Annual Review of Anthropology*, 28, pp. 201–224.

Murray, R.J., 2001. "The man that says slaves be quite happy in slavery… is either ignorant or a lying person…" An Account of Slavery in the Marginal Colonies of the British West Indies. Unpublished PhD thesis, University of Glasgow.

Nellis, D.W. and Everard, C.O.R., 1983. The biology of the mongoose in the Caribbean. *Studies on the Fauna of Curacao and Other Caribbean Islands*, 64(1), pp. 1–162.

Pimm, S., Raven, P., Peterson, A., Şekercioğlu, Ç.H. and Ehrlich, P.R., 2006. Human impacts on the rates of recent, present, and future bird extinctions. *Proceedings of the National Academy of Sciences*, 103(29), pp. 10941–10946.

Perdikaris, S., Bain, A., Grouard, S., Baker, K., Gonzalez, E., Hoelzel, A.R., Miller, H., Persaud, R. and Sykes, N., 2018. From icon of empire to national emblem: New evidence for the fallow deer of Barbuda. *Environmental Archaeology*, 23(1), pp. 47–55.

Perdikaris, S., Boger, R. and Ibrahimpasic, E., 2021a. Seduction, promises and the Disneyfication of Barbuda post Irma. *Translocal Contemporary Local and Urban Cultures Journal*. Number 5 (un)inhabited spaces. Ana Salgueiro and Nuno Marques (eds). ISSN 2184-1519 Madeira and Umeå. Translocal.cm-funchal.pt/2019/05/02/revista05/

Perdikaris, S., Boger, R., Gonzalez, E., Ibrahimpašić, E. and Adams, J., 2021b. Disrupted identities and forced nomads: A post-disaster legacy of neocolonialism in the island of Barbuda, Lesser Antilles. *Island Studies Journal*, 16(1):pp. 115–134.

Perdikaris, S., Grouard, S., Hambrecht, G., Hicks, M., Mebane Cruz, A. and Peraud, R., 2013. The caves of Barbuda's eastern coast: Long term occupation, ethnohistory and ritual. *Caribbean Connections*, 3, pp. 1–9.

Preston, C., 2009. The terms 'native' and 'alien' – a biogeographical perspective. *Progress in Human Geography*. https://doi.org/10.1177/0309132509105002.

Schlaepfer, M.A., Sax, D.F. and Olden, J.D., 2011. The potential conservation value of non-native species. *Conservation Biology*, 25(3), pp. 428–437.

Shackelford, N., Hobbs, R.J., Heller, N.E., Hallett, L.M. and Seastedt, T.R., 2013. Finding a middle-ground: The native/non-native debate. *Biological Conservation*, 158, pp. 55–62.

Shine, R., 2010. The ecological impact of invasive cane toads (Bufo marinus) in Australia. *The Quarterly Review of Biology*, 85(3), pp. 253–291.

Skandrani, Z., Lepetz, S. and Prévot-Julliard, A.C., 2014. Nuisance species: Beyond the ecological perspective. *Ecological Processes*, 3(1), pp. 1–12.

Smith, S.R., van der Burg, W.J., Debrot, A.O., van Buurt, G. and de Freitas, J.A., 2014. *Key elements towards a joint invasive alien species strategy for the Dutch Caribbean* (No. C020/14). IMARES/PRI.

Sykes, N.J., White, J., Hayes, T.E. and Palmer, M.R., 2006. Tracking animals using strontium isotopes in teeth: The role of fallow deer (Dama dama) in Roman Britain. *Antiquity*, 80(310), pp. 948–959.

Ünala, Y. and Çulhacıa, H., 2018. Investigation of fallow deer (Cervus dama L.) population densities by camera trap method in Antalya Düzlerçamı Eşenadası Breeding Station. *Turkish Journal of Forestry*, 19(1), pp. 57–62.

Valenzuela, A., Baker, K., Carden, R.F., Evans, J., Higham, T., Hoelzel, A.R., Lamb, A., Madgwick, R., Miller, H., Alcover, J.A. and Cau, M.Á., 2016. Both introduced and extinct: The fallow deer of Roman Mallorca. *Journal of Archaeological Science: Reports*, 9, pp. 168–177.

Van der Burg, W.J., De Freitas, J., Debrot, A.O. and Lotz, L.A.P., 2012. *Naturalised and invasive alien plant species in the Caribbean Netherlands: Status, distribution, threats, priorities and recommendations: Report of a joint Imares/Carmabi/PRI project financed by the Dutch ministry of economic affairs, agriculture & innovation* (No. C185/11). Plant Research International, Business Unit Agrosystems Research.

Warren, C.R., 2007. Perspectives on the 'alien' versus 'native' species debate: A critique of concepts, language and practice. *Progress in Human Geography*, 31, pp. 427–446.

Woinarski, J.C., Burbidge, A.A. and Harrison, P.L., 2015. Ongoing unraveling of a continental fauna: Decline and extinction of Australian mammals since European settlement. *Proceedings of the National Academy of Sciences*, 112(15) pp. 4531–4540.

Woinarski, J.C., Legge, S., Fitzsimons, J.A., Traill, B.J., Burbidge, A.A., Fisher, A., Firth, R.S., Gordon, I.J., Griffiths, A.D., Johnson, C.N. and McKenzie, N.L., 2011. The disappearing mammal fauna of Northern Australia: Context, cause, and response. *Conservation Letters*, 4(3), pp. 192–201.

Wyatt, K.B., Campos, P.F., Gilbert, M.T.P., Kolokotronis, S.O., Hynes, W.H., DeSalle, R., Daszak, P., MacPhee, R.D. and Greenwood, A.D., 2008. Historical mammal extinction on Christmas island (Indian Ocean) correlates with introduced infectious disease. *PloS one*, 3(11), p. e3602.

Zeuner, F.E., 1963. *A history of domesticated animals*. London: Hutchinson of London.

5 From the far ground to the near ground

Barbuda's shifting agricultural practices

Amy E. Potter

5.1 Introduction

The island of Barbuda, part of the twin-island country of Antigua and Barbuda in the Lesser Antilles Caribbean, has undergone a tremendous transformation in its economy over the last several decades. Islanders have transitioned from largely subsistence activities, small-scale food export grown on the island's commons, and livestock herding, to a greater dependence on food imports – and an economy increasingly centered on lobster export and touristic endeavors (Potter and Sluyter 2010; 2012; Potter 2011, 2015; Potter et al. 2017). In addition, Barbudans also must contend with larger forces at work, particularly the vulnerabilities of global climate change and concerns over food security (Boger et al. 2016). More recently, islanders are rebuilding after Category 5 hurricane caused considerable damage to their infrastructure in September of 2017. Additionally, given the ways Barbuda's 'disasterscape' has been exploited after the hurricane, it is critical to understand how Barbudan's unique culture and identity, particularly as it relates to agricultural practices, are under threat (Perdikaris et al. 2020a & 2020b).

Over the last decade, there has been a growing movement within the Caribbean, a call to seriously address the region's reliance on food imports with a trend for island-states to move toward more agriculture independence (Cave 2013). A number of scholars have sought to understand food security within the region in terms of contextualizing the problem amongst the immediate economic crisis and challenges posed by climate change (Beckford and Campbell 2013; Tandon 2013; FAO 2012). Beckford and Campbell write, 'Food security in the Caribbean must be analyzed in the context of the realities of most of the countries in the region' (2013: 33). It is within this context that I seek to highlight the unique realities facing the island of Barbuda against the backdrop of these larger regional problems of climate change and food security based on fieldwork conducted before Hurricane Irma. Building on previous work on the livestock practices in Barbuda and preliminary research on backyard gardening as a form of resilience (Potter and Sluyter 2010; Boger et al. 2014; Boger et al. 2016; Perdikaris et al. 2013,

DOI: 10.4324/9781003347996-6

2014), this chapter will center exclusively on the changing agricultural sector in Barbuda, with a focus on Barbuda's backyard gardens.

In order to do this, the chapter will briefly examine the literature on backyard gardens and the growing threat of food insecurity in the Caribbean, and then describe the author's methods. The chapter will then shift to the history of Barbuda while emphasizing changing agricultural practices on the island. The author will then spend the remainder of the chapter describing Barbudan backyard gardens before Hurricane Irma as revealed by the household survey and semi-structured interviews followed by conclusions.

5.2 Background literature on backyard gardens and food security

The study of gardens has a deeply rooted history in cultural geography (Sauer 1952; Kimber 1966, 1973).

> Those of us who have worked extensively in the tropics, whether in the New World or the Old World, know that house gardens are a widespread and important element in the domestic economy of rural communities, but because they are relatively inconspicuous and less visually impressive than field systems, they tend to be overlooked and their contribution to subsistence underrated.
>
> (Whitmore and Turner 2001, vii–viii)

Carissa Kimber's early work on the Dooryard Gardens of Martinique (1966) described the gardens as managed artifacts where gardeners both curated desired plants and took advantage of spontaneous germination. A few years later her research on gardens in Puerto Rico (1973) produced a typology of garden designs that paved the way for a robust area of research. In 2004, nearly 40 years after her piece on Martinique, she described in a special issue in the *Geographical Review* how cultural geographers have broadened the study of gardens to also examine their spatial characteristics and biological content, to consider garden plants as cultural traits, and to explore garden plants as design elements. 'These studies suggest the validity of local Caribbean farmers' traditional agricultural knowledge, and sustainable-agriculture investigators recognized the role that house gardens or kitchen gardens could play in the sustainable-agriculture programs for the islands' (Kimber 2004, 270).

More recently in the Caribbean, gardens are increasingly viewed as an important solution to combat food security concerns in the region.[1] In August 2013, a *New York Times* article described Jamaica's return to agriculture. The author Damien Cave writes:

> Across the Caribbean, food imports have become a budget-busting problem, prompting one of the world's most fertile regions to reclaim

its agricultural past. But instead of turning to big agribusinesses, officials are recruiting everyone they can to combat the cost of imports, which have roughly doubled in price over the past decade. In Jamaica, Haiti, the Bahamas and elsewhere, local farm-to-table production is not a restaurant sales pitch; it is a government motto.

Jamaican officials also gave out seed kits to encourage backyard gardening.

Beckford and Campbell argue that while food security has not been problematic in the region in the past, it is increasingly coming to the forefront for Caribbean leaders as they realize the potential impacts of climate change, natural hazards, declining local food production, and dependence on imported foods, and growing preference for North American food convenience has changed the region (2013). This dependence on imported food largely from North America has also changed local diets and has led to an increase in diet-related diseases. Caribbean people are generally moving away from local foodways. The region has seen an overall decline in land productivity, labor, and management. Earnings from traditional export crops have also declined, which has resulted in an inability to purchase food. Caribbean markets have lost their trade preferences for traditional export crops. Poverty in the region is increasing, which also impacts people's access to food (Beckford and Campbell 2013; Potter et al. 2017).

Beckford and Campbell have also noted that these gardens[2] generally used for home consumption and sharing are 'typically small cultivated areas in the backyard, where vegetables ... and other short-term crops are grown' (Beckford and Campbell 2013: 95). They are seen as one solution to the growing threat to food security in the region. Gardens serve a variety of important functions in that they are (1) sites of agricultural training and experimentation, (2) provide household food security, playing a vital role in nutrition and calories, (3) showcase traditional agricultural knowledge, and (4) are valuable sources of income (Beckford and Campbell 2013: 106–107).

In Barbuda's sister island of Antigua, there has been some urgency to address similar concerns relating to declining agriculture in the region. According to 2011 data, only 3.8% of Antigua and Barbuda's GDP centered on Agriculture compared to 32.9% on Industry and 51.8 on Services. The Antigua Ministry of Agriculture cited in 2010 that the twin islands imported approximately 16 million pounds of vegetables. In an effort to reduce those imports by half, Antigua launched a program in 2011 to address issues relating to food security and put forth efforts to encourage their population to produce food annually in their backyards, with a larger goal of producing locally four million pounds of vegetables (Antigua and Barbuda 2012; Gordon 2012; 2013). Every May, Antigua government officials recognize National Backyard Gardening Day in an effort to promote citizen participation in local food production.

While we know generally what is happening in Antigua related to food security, this same information cannot simply be extrapolated to Barbuda.

Food security and climate change as it relates to Barbuda must be understood against the backdrop of the island's unique history, culture, land tenure, and geography that set it apart from the larger Caribbean region. This chapter will build upon previously published preliminary research on the Barbudan backyard garden (Boger et al. 2014; Boger et al. 2016; Perdikaris et al. 2013, 2014).

5.3 Methods

Utilizing long-term ethnographic fieldwork, the author employed a mixed-methods approach (Crang and Cook 2007), which included semi-structured interviews[3] conducted between 2007 and 2013, participant observation of farmers, as well as a survey of 33 households conducted in July and December of 2013 (Berleant-Schiller and Pulsipher 1986).[4]

5.4 Background on Barbuda

Barbuda, with a population of 1,634 people (Antigua and Barbuda Census 2011) on 62 square miles, was leased for nearly two centuries (1680–1870) by the Codrington family of England who used the enslaved African population to grow provisions and raise livestock for sugar plantations on neighboring Antigua. Unlike its Caribbean counterparts, the island was not suited to large-scale agriculture due to its arid climate and relatively thin soils (Harris 1965; Berleant-Schiller 1977, 1978). Codrington's repeated attempts to establish plantations for cotton and other crops failed, in part, due to the shallow soils, an average annual rainfall of only 35.4 inches, a long winter-dry season, frequent droughts, and a scarcity of surface water (Berleant-Schiller 1977).

   Over time, Barbudans developed a complex herding ecology that allowed them to adapt to periods of drought and abundant rainfall. Barbudans successfully navigated these droughts by largely abandoning commercial cultivation during dry years and expanding stock keeping, in essence creating a complex herding ecology to navigate the annual rainfall variation (Berleant-Schiller 1983). During dry periods, livestock would assemble at the wells drawn particularly to the watering trough where Barbudans would pen the cattle. During wet periods, the communal land tenure allowed livestock to forage, breed unrestricted, and drink freely from the wells and sinkholes where water collected. When needed, Barbudans would round them up for butcher and export. It was also during wet years that Barbudans resumed crop production and restricted the movement of livestock with fencing. Berleant-Schiller writes of this symbiotic relationship: 'Land use and tenure together have preserved the community from the hazards of drought and domination' (1983, 87).

   Enslaved Africans overcame the island's environmental constraints (drought and thin soils) and succeeded in creating sustainable small-scale agriculture for hundreds of years, whereas Europeans were unable to overcome those same environmental constraints to develop large-scale plantations on the island as in other parts of the Caribbean. Two Barbudans described the old (and in many ways current) planting methods in this way:

> In dat time gone by you have fi take the cutlass and you have fi cut down all the small bushes. And after you done do dat dere, you would have fi take the ax and cut the bigger trees, it take you time, plenty time to get everything, then after that, when you finish that now. You have fi burn up all those weeds. You understand. Den after you done burn up the bushes and so, you have to fence up the, you have fi enclose the whole place ... When you finish, do that. No animal can get in, you understand what I mean. (2012)

> What they used to do we call it slot. Slot. Cut the ground and they burn it so that now keep off all the insect from bother the plant and them. So when you cut the ground and you burn it you left it to cool and after you plant it. You fence your ground. We used to call it barricada. (2012)

Over time, this shifting cultivation gave way to a successful vegetable export economy that thrived into the 1980s and 1990s to nearby islands. One Barbudan Emmanuel Lewis recalled in an interview,

> We used to grow ground provisions, and export them to the neighbouring islands. I used to go to St. Martin, St. Kitts, Nevis and Anguilla on my boat selling sweet potato, yam, cassava and pumpkin, and at a young age too.
>
> (Cornelius 2007, 108)

5.5 Shifting agriculture practices

This successful herding ecology and economy of agriculture have slowly waned in Barbuda as in other parts of the region (see Potter and Sluyter 2010). This section will explore the complexity of factors involved in agriculture's decline.[5]

In the 1960s, alternative forms of wage labor came in the form of a new hotel at Coco Point. Other hotels over time followed. For the first time, Barbudans could find employment where they were no longer tied directly to Barbuda's commons (both land and sea).

In addition to hotel employment, since 1975, a succession of foreign companies quarried sand at Palmetto Point for use in construction projects throughout the Caribbean (Coram 1993; deAlbuquerque and McElroy 1995). Those sand exports have resulted in jobs for Barbudans, royalties for the federal and local governments, and detrimental environmental impacts such as groundwater contamination (Campbell 2006). The Barbuda Council, the locally elected body established in 1976 during the period of associated statehood, benefitted from these very royalties and in the past, was able to use the revenue to offset their payroll of an estimated 400 workers at an average annual salary of $17,000 XCD, equivalent to $6,513 USD (Potter and Sluyter 2010).

Remittances from Barbudans living abroad are another major source of income. By some estimates, three times as many Barbudans live in New York City as on Barbuda, with additional communities in other parts of the United States such as the Virgin Islands, the United Kingdom, and Canada (Potter 2011). No comprehensive statistics exist on either the numbers of Barbudans living abroad or on the amount of money they remit to the island, but a census dating to 1971 revealed that more than half of the households received some form of support from abroad, including cash or gifts sent by relatives in New York City or elsewhere, pension or social security checks, and wages from seasonal employment abroad (Rive Berleant-Schiller Papers, Various papers, n.d). Remittances and seasonal migratory employment abroad certainly continue and might have grown.

Barbudan migration is one of the major catalysts for undoing agricultural land use in the far ground, traditional land-use plots scattered throughout the island. As one Barbudan return migrant told me:

> People stop, the older people who used to cultivate the land they themselves have gotten too old and tired to do it. The children left. They sent them away. As soon as they finished school and they can afford it they send them to England or Canada. They send them away and who is left to do cultivation? They're already old and tired. (2009)

Another Barbudan described it in this way:

> Persons migrate a lot and also people who used to grow a lot. They just go away and sometimes I would say and I'm not afraid to say probably because we left for such a long time. Things might have because people did not understand how it was before. I hardly have anyone in my age group here. More of them are younger or the few are older cannot move around too bad to do anything so we just wonder if it's because we had left. And that vacuum was never filled. (2012)

It was also Barbudan migrants who helped change consumptive food practices by introducing relatives to foreign products sent to the island by way of barrels, which eventually paved the way for grocery stores selling imported food items (see also Lowenthal and Clarke 1982). One Barbudan migrant's grocery store sprung up organically after she started sending food items to her mother.

So stuff I sent to my mother all the time, I sent her big boxes of food, turn into barrels sometimes of food ... I would send her coffee creamer and some people would say 'Oh can you get me a bottle of that.' And then I would send her, she likes that Folgers Dry Roasted Coffee, she just liked the aroma ... and I would send extras so that's how I got into you know [the grocery store business]. (2010)

Some Barbudans in recent years have expressed frustration over the quantity of imports, particularly the importation of foodstuffs Barbudan fore parents used to grow.

I'd like to see turn a good quarter portion of this land into just agriculture lands for someone who can develop, work the land, grow and feed the world. Feed us. Feed the nation. We used to do that. Our parents used to do it. They would grow what we can't eat we'd export it. And now we're importing things we used to export ourselves. (2010)

It is this same frustration over imported food items that has served as an impetus to return to gardening, albeit on a much smaller scale.

In 2013 at the time of the survey, 80% of the island's food came through weekly deliveries via refrigerated containers and from Antigua to Barbuda via a supply boat. The problem of economic dependency (food from outside is terribly expensive) is its contribution to the creation of poverty but also to increased malnutrition and related disease through the consumption of processed foods and a shift away from traditional foods and nutrition (Beckford and Campbell 2013). Another problematic factor with imported foods and reliance on refrigerated containers and a supply boat for sustenance is that the supply is unreliable in times of bad weather. During Hurricane Lewis in 1995, Barbuda was without communication with the outside world and was without any food and supplies from the mainland for over a month (Boger et al. 2016).[6]

The economic processes and migration of Barbudans are not enough to explain the demise of agriculture. There are also negative cultural associations. One Barbudan in her mid-1920s told me in 2007,

I don't know why it changed, it seemed like it was better off for us. I think it changed because the Council started hiring a lot of people and agriculture was seen as for poorer people and stupid people. And then I guess we just lost our vision. I can remember in our yard we used to grow peas and grow produce. Stuff like that. (2007)

The Barbudan youth's disinterest in agriculture could also be the result of a Council policy that punished workers by sending them to labor at the Agricultural Station. One Barbudan woman said the Barbuda Council is sending a message that says, 'You're no good and agriculture is not good enough so that's the only place we can send you ... In my opinion I cannot see anyone would look at this as a positive thing to do' (Hart 2010: 26).

As the economy has changed, so too have the uses of Barbuda's commons. Lands that were formerly devoted to agriculture and livestock are now allotted for home construction as the village of Codrington expands. In 1982, the village wall enclosing the village was torn down. Its complete removal signaled the official sanctioned expansion of the village outside its confines and the symbolic disintegration of the Barbudan community. While the wall enclosing the village was torn down in 1982, the beginning stages of increased residential land use and the expansion of the village were documented on the pages of the *Barbuda Voice* as early as 1970 as Barbudans sought to acquire land, albeit often unsuccessfully, outside the traditional confines of the Codrington village. Prior to the formation of the Barbuda Council in 1976, an island warden apportioned the land. Records dating back to the formation of the Barbuda Council and its subsequent 1977 takeover of land allocation show an increasing interest in acquiring land for home construction or business endeavors (Potter 2011).

In the midst of these land-use changes, Barbudans formalized into law their communal land tenure through the Barbuda Land Act, 2007, in January of 2008 (Antigua and Barbuda 2008).[7] However, it was agricultural (and livestock) activities that gave Barbudans the grounds to contend that they owned the island in common. In the wake of these changes, the backyard garden or near ground has taken on increased significance as never before in the island's history. Barbudan elder Papa Joe described the difference between the traditional far ground plots to the Barbudan near-ground or backyard garden in 2012.

> Well the near ground is supposed to, well okay, the one you would call near ground which your yard, the yard you live in, you farm your yard ... where I live, I'm doing some work on it. Well time gone by when I was younger, I would still do this where I live and I would go out in the forest and do one out there, but today now, I'm old man now, I can't do that ... I have to stay here, you know because I can't walk to go out in the forest like I used to do.

The remainder of this chapter will explore the Barbudan backyard garden or near ground.

5.6 Barbuda's backyard garden or near ground

In July and December 2013, I conducted a survey of 33 Barbudan households modeled in part after Riva Berleant-Schiller and Lydia Pulsipher's survey of subsistence cultivation plots on the islands of Barbuda and Montserrat (1986). The survey was initiated in the summer months as many Barbudans were preparing their gardens for the rainy season and then again at the height of the growing season – the primary growing season is September through January. Most of the surveys were administered at the home of the survey taker and often included a tour of the garden. Table 5.1 provides a breakdown of the household demographics. The surveys represented 102 people (roughly 6% of the Barbudan population), with an average household size of 3.1. The household size ranged from a single individual to 15 people. The average age of the survey respondent was 56 (ranging from 19 to 82). Eighteen females and fifteen males filled out the survey. Most of the surveys were completed by retirees; however, some identified as teachers, nurses, and farmers. Many of the gardens are tended by the individual who completed the survey, with others receiving additional help from a partner or children. Nearly half of the gardens included in the household survey were located in the Park section of Codrington village.

Table 5.2 provides a list of plants grown in approximately 25% (8 of 33) of household gardens in 2013. Like Berleant-Schiller and Pulsipher (1986), my aim was to examine crops and gardening techniques grown in the near ground rather than productivity and yield (p. 6). In addition to fruit trees, vegetables, legumes, and tubers, most gardeners also reported growing herbs, teas, spices, and medicinal plants, which included aloe, balsam, mint, sage, fever grass, lemon grass, and bitter bush (for colds), as well as French thyme.

Table 5.3 provides a list of gardening techniques and land management systems. Many gardeners I spoke with adhered to traditional planting knowledge passed down from their fore parents. In interviews with several gardeners, I heard the same refrain, 'I plant by the moon', a method they had observed from a parent, particularly when it comes to planting certain crops like pigeon peas.

Table 5.1 Household survey demographics

Demographics	
Total surveys	N = 33
Gender	
Male	45%
Female	55%
Average age	56
Youngest participant	19
Oldest participant	82
Average household size	3.1
Household location*	
Park	45%
Middlesect	9%
Mulatto	21%
Other	24%

Due to rounding, not all categories may add up to 100%.

Table 5.2 Inventory of plants in 25% of backyard gardens

Fruit trees	(%)	Fruits and vegetables	(%)
Avocado	33	Chives	27
Banana	61	Hot pepper	24
Coconut	36	Maize/corn	45
Finger roll	39	Okra	55
Genip	24	Pumpkin	64
Guava	36	Spinach	58
Lemon	39	Squash	48
Lime	48	Sweet pepper	76
Mango	58	Tomato	64
Papaya	52	Watermelon	36
Plantain	36		
Pomegranate	33	**Legumes/tubers**	
Sugar apple	55	Bean	30
		Black eyed pea	27
Other		Peanuts	30
Sugar cane	24	Sweet potato	55
		Tan up a heap	33

Table 5.3 Planting and harvesting techniques

Planting and harvesting techniques	Yes (%)	No (%)
Strategic plant placement	70	30
Continuous harvesting	67	33
Cuttings saved	42	68
Seeds saved	64	36
Land management	**Yes (%)**	**No (%)**
Mulch	24	76
Sand	61	39
Burning	21	79
Fertilizer	61	39
Land rotation	36	64
Crop rotation	48	52
Pest control	64	36

As noted by Berleant-Schiller and Pulsipher in their study of Barbuda (1986), gardening techniques that were practiced in the 'far grounds' have been maintained in the 'near ground' such as intercropping, crop rotation, and forking the soil. Of the gardeners who use fertilizer, 60% use animal manure from donkeys or cows left on the road. Only two gardeners surveyed buy fertilizer from the store due to cost inhibitors. Sixty-four percent of gardeners saved their seeds and 42% saved cuttings.

The survey revealed that Barbudans used a variety of measures to control pests. A gardener who lives outside the confines of the village described the benefits of burning: 'Well because of the insects in the ground. You know when you burn it. It's better for the plant. They grow better'. Burning is

Table 5.4 Water use

Water use	Yes (%)	No (%)
Add water to garden?	97	3
Water sources		
Cistern	42	58
Black tank	48	52
Drum	48	52
Well	42	58
Government water	61	39
APUA water line	61	39

a bit trickier for gardeners who live within the confines of the village of Codrington, as burning is not allowed. Other methods include purchasing a chemical spray from Antigua (Sevin) to control worms. Others will use a combination of neem powder mixed with water and a household cleaner (Squeezy).

Table 5.4 details the water sources by household (see also Boger et al. 2014 and Boger et al. 2016 for further discussion). Ninety-seven percent of households reported adding water to their gardens. Backyard garden agriculture is highly dependent on additional water outside of rainfall, which is of particular concern given future projections of rainfall totals in the region.[8] One female gardener cited the benefit of her backyard garden over the far ground agriculture in this way: 'Backyard gardens allow you to grow even when there is not rain' (2013).

However, not all Barbudan water should be used in the garden. Boger et al. 2016 found that water samples collected from government water and wells in 2014 and 2015 had salinity ranges from 0.25 to 10 ppt. Of the 61% of households that had access to government water, only 36% used government water on their garden crops, with almost all specifying that they were careful not to put water on the leaves because of salinity levels. In addition to careful use of government water, in order to cope with the salinity levels of both well water and government water, one 60-year-old gardener put in a black tank and catches additional rainwater with buckets, 'That's how much me like agriculture me get a tank for the garden. When the rain fall, we try to save the good water' (2012).[9] One gardener I interviewed did not realize the salinity levels of the water she sprayed on her garden that had the effect of yellowing or killing the plants altogether. She told me in July of 2013, 'But now I do know that it more or less kills the plant'.

Not all islanders, however, can afford the cost of a black tank for their garden. One backyard gardener has instead created a rock-lined canal in his backyard to catch rainwater. The extensive garden makes use of every available inch of space in his yard for rainwater collection, raised fields, and an irrigation canal. Other creative adaptations include the use of plastic sacks to hold moisture in the soil during the dry season.

Table 5.5 Yard animals

Animals	Yes (%)	No (%)
Chickens	9	91
Fowl	12	88
Land crab	48	5
Land turtle	55	45
Sheep/goats	6	94

Table 5.6 Subsistence use or surplus

Subsistence use or surplus	Yes	No	Other	
Surplus	82%	12%	6%	
	Giveaway	Sell	Combination giveaway and sell	N/A
What do you do with surplus?	18%	27%	36%	18%

The 'Other' category reflected those gardeners who had recently begun gardening.

Table 5.5 is a survey of animals kept in the yard. As other scholars have noted, gardens are important spaces 'to raise small animals on food scraps' (Christie 2008, 35). Of particular interest in terms of kept animals is that greater than half of households surveyed had land turtles within the yard, while nearly half of households had land crabs. The preparation of turtles as part of the weekly Sunday meal is a longstanding cultural tradition on the island as one Barbudan described, 'Well I love to go to the woods, I go hunting for turtles, because as you know, we eat them. Yeah it's a tradition, we do eat them' (2013). Another Barbudan female gardener elaborated on their use as pets as well as their medicinal benefits:

> We raise them because we eat them, also they are nice pets. You can just feed them. We have a big pen and we backed it around with stone, and things that are left on the ground. Like if you have a melon spoil or we don't have any rain, then we can just pick them up and feed the turtle or feed the crab ... You would love it, they say it is good for your body you know. Especially for people who are sick and down. They get the smaller one and they boil like a broth, strain it and they give them the water to drink. (2013)

Barbudan gardens serve as important spaces 'to obtain ingredients for traditional cuisine, to maintain cultural inheritance, and to preserve embodied knowledge and reciprocity networks' (Kimber 2004, 272). In addition to turtles and crabs, 12% of households had fowl and less than 10% kept chickens, sheep, and goats. Three of the respondents usually had sheep and goats (but not at the time of the survey), while one respondent had sheep and goats out on Goat Island.

Table 5.6 provides details on whether households experience a garden surplus and if the household garden provides additional income. Eighty-two percent of households reported a surplus from their gardens. One gardener sold his surplus to an island grocery store and reported additional earnings of $300 ECD.

> But I think it has become significant. Feed yourself, then you are practically better off then most, because when my garden is in full swing it saves me a lot because a lot of stuff I don't have to buy and then I have surplus to sell. So when it is in full swing it helps me out a lot. (2013)

Other studies have shown how gardens are important components of reciprocity networks (Kimber 2004). At least 18% of those who experienced a surplus gave away items from their garden, with 36% of gardeners engaged in both selling and giving away of their garden surplus. During one of my interviews with a gardener in 2013, I witnessed this reciprocity network in action as a gardener received fresh catch from a local fisherman in exchange for corn and tomatoes from his garden.

Similar to the Berleant-Schiller and Pulsipher's (1986) study, I mapped two Barbudan backyard gardens to illustrate the contrast between gardens in the traditional confines of Codrington and those on lands outside the village.

Garden 1 (see Figure 5.1) is located in the Park area of Codrington, where homes and plots of land are generally smaller. The gardener was a

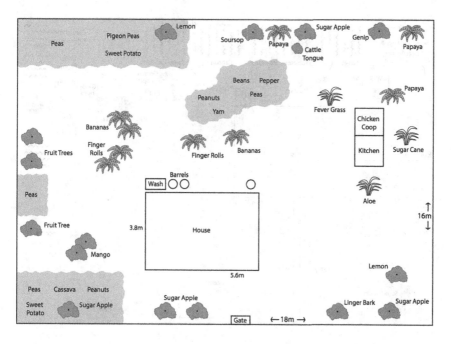

Figure 5.1 Map of backyard garden in park, Codrington Village. Map by Marissa Alessi, University of Mary Washington Geography.

75-year-old retired male, who tended his garden alone. The fenced area of his yard measured roughly 288 m², with the home built in the 1950s measuring 21.28 m². A casual observer would not recognize the complexity and care of this carefully planned garden filled with peas, beans, and a variety of fruit trees. 'Caribbean kitchen gardens demonstrate a highly intensive use of land space with a profusion of crops and other plant species in a pattern seemingly without order' (Beckford and Campbell 2013, 99). This gardener grows plants throughout the year, mixing soil from the Highlands with sand. He uses barrels and buckets to collect rainwater to water his plants, as illustrated in Figure 5.1.

Garden 2 (Figure 5.2) is located outside the traditional confines of Codrington in an area called Big Pond. The primary gardener was a 38-year-old Dominican male in a household of four people. The family had lived in their home for two years. Prior to this time, they had previously lived in the village, but it was there he had trouble growing food amongst the rocky soil and limited space. Their yard in the village was not fenced, and animals would walk through the garden. The size of their new property is roughly 1,600 m² with 130 m² house. Compared to the first garden (Figure 5.1), you see the use of intercropping as well as ornamental plants. The front part of the yard was rocky with thin soil, which is why the gardener placed ornamental plants in the front. For water, this gardener relies

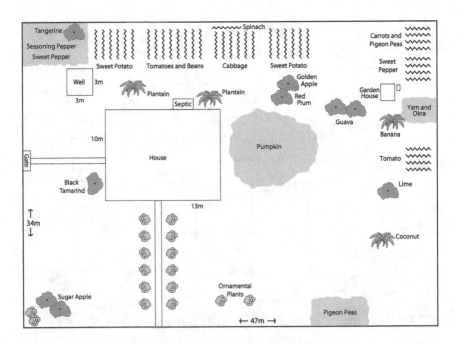

Figure 5.2 Map of backyard garden in big pond outside of Codrington. Map by Marissa Alessi, University of Mary Washington Geography.

on a well, a black tank, and drums. This gardener uses a hoe, pick, and axe and relies on the technique of banking (a technique he learned from his father in Dominica), which allow him to drop a hose between banks to irrigate the plants in the evening and protect the plants from the pooling of water in the yard.

5.7 Challenges for backyard gardeners

Barbudan backyard gardeners face a number of constraints similar to other home gardeners documented in the scholarly literature (Gahlhena et al. 2013). Despite efforts in Antigua to promote backyard gardening – the central government launched an initiative in 2011 – Barbudans in 2012 and 2013 expressed concern that little seemed to be trickling over to the sister island. Backyard gardeners complained of their limited access to seeds and their struggle for equipment to clear grounds for planting. Yet despite these obstacles, gardeners are saving seeds, sharing seeds, and creatively navigating the environmental constraints around them. There have been a number of initiatives from Barbudans and overseas agencies that have sought to encourage farming, including USAID.

Barbuda is a karst landscape, with shallow, rocky soils that exhibit poor drainage and high salinity. Expansion outside the village has brought forth settlement growth in low-lying areas. Several Barbudans, in order to build homes and plant crops, have had to transport soil from the Highlands to build up the land around.

> You know something, this whole area here, was a pond. When I say pond, heavy rain it holds water. So I have to bring in sand and fill it up. When I get a good amount of rain water stay holds. Goes out comes off and comes in. Sometimes I have peanuts also in here. But one time I had a lot of rain and everybody drowned. That's why I turning to the planting of because if you get rain, well the water wouldn't trouble these things. I forget about the peanuts and put in these things right here. (2013)

Kimber reminds us too that gardens can reflect larger societal problems (2004). Barbuda's changing economy also parallels changing ideas of community and eroding social cohesion that has had detrimental effects on agriculture (Berleant-Schiller 1991). This social fragmentation has at times produced a careless disregard for others' property, particularly as it relates to agriculture. Several Barbudans within their interviews noted the destruction of their crops when not in the confines of their yard. One Barbudan woman in her 1950s stopped cultivating completely in the far grounds after vandals destroyed her crop.

Well I used to cultivate only further out. But I stopped. The reason for it is that people used to go in the ground and take things, and I don't like it ... And I remember one time it was down on the River area I have a ground it's bigger than this yard and I'm telling you, I used to go down every day to it. I would go early in the morning, come home and cook lunch and at two, three o'clock back again until at night. And I go to my ground one day just about two, take out bunch of some corn that I had when I go I cry. Somebody break off all and throw them on the ground because they wasn't ripe. And I stand up and I cry and I said, 'That's it for me.' But I try to encourage myself in the yard and here's where I do all of my sort. (2012)

The fenced backyard is a way to ensure the protection of crops, fruit trees, and the effort put into growing. As one 49-year-old male Barbudan noted of his father who relocated his crops to his backyard, 'Can see his stuff so no one can touch it'.

5.8 Conclusions

The island of Barbuda has a longstanding tradition of agricultural practices that have adapted and changed over the last several decades. This chapter has described these transformations and provided an examination of the key characteristics of the island's backyard gardens prior to Hurricane Irma in 2017. Regrettably, however, the sinister work of disaster capitalism and environmental destruction taking place on the island after Hurricane Irma are all working in tandem to undermine Barbuda's unique cultural practices and identity. 'The race for profit' threatens Barbudans 'tangible and intangible cultural heritage' (Perdikaris et al. 2020a, 11; Perdikaris et al. 2020b).

For example, in the months following Hurricane Irma, a number of plans surfaced under the guise of 'helping' Barbudan people impacted by the hurricane return home. One such plan, from a Chinese firm, proposed constructing 'high-rise apartment buildings', an unusual housing structure for an island that boasts nearly 68% of households owning their home outright (Antigua and Barbuda Census 2011).

'The complex would be self-contained, meaning that it would contain a supermarket, cleaners, a drug store, a restaurant, a movie theater, clothing stores, a church, car parks and other kinds of amenities that are common to new complexes,' The Cabinet notes by Chief of Staff Lionel 'Max' Hurst read.

(Antiguanewsroom.com 2017)

The Chinese company, in its efforts to secure a contract to build a housing structure typical of China's already large and rapidly growing cities, failed to consider housing and communal property traditions as well as cultural practices, one of which is the planting of food in the house yard. Anthropologist Anthony Oliver-Smith has drawn attention to resettlement plans like the one cited above, where communities impacted by a variety of disasters endure the onslaught of projects that hinder the rebuilding of the community. He writes that after disasters 'people must struggle to construct a life world that clearly articulates their continuity and identity as a people again' (2005, 48) and that 'such projects are about reconstructing communities after they have been materially destroyed and socially traumatized to varying degrees' (2005, 55).

Based on interviews and household surveys, Barbuda's backyard gardens or near grounds are not only important sources of food and supplemental income, but they are also materially and socially important to the island's community (Oliver-Smith 2005). They are places where traditional knowledge is practiced and passed down and spaces where cultural traditions persist. The surplus of agriculture additionally builds community through reciprocity networks amongst island neighbors. As the island and larger Caribbean region works toward increased food security and greater resiliency, this author calls for more research and greater attention to Barbuda's backyard gardens, especially now. Michael Woods writes,

> Not only will climate change itself have direct impacts on the rural environment, and hence on rural economies and societies, but the responses and adaptation of rural communities to climate change will determine their future social and economic viability, while harnessing renewable rural resources could also play a key role in mitigating and limiting the effects of climate change.
>
> (2012, 128–129)

The backyard garden will most certainly be a key component to moving forward.

Acknowledgments

This research was made possible in part by a faculty development travel grant from Armstrong Atlantic State University in 2013. I would like to thank Khristina Williams, Sophia Perdikaris, Rebecca Boger, Jennifer Adams, Katie Hejtmanek, John Mussington, and the students of the 2012 Brooklyn College GIS Field School in Barbuda. This research would not have been possible without the expertise and knowledge of Barbudan backyard gardeners.

Notes

1 Food security is centered on the premise 'that all people at all times have access (including physical, social and economic) to sufficient, safe and nutritious food necessary to lead active and healthy lives' (FAO 2009, quoted in McDonald 2010: 2).

2 These gardens have a variety of names within the academic literature, sometimes called home gardens, kitchen gardens, dooryard gardens, kitchen space.
3 The author will provide the year when the interview was conducted and maintain the anonymity of most interviewees. All interviews in this chapter were conducted in Barbuda between 2007 and 2013.
4 Potter conducted fieldwork in Barbuda between 2007 and 2013 (see dissertation Potter 2011).
5 The island of Barbuda is not completely void of agriculture, and the author wants to be careful not to present it as such. The author observed in 2012/2013 that there are three types of agriculture taking place: farming of council lands, small-scale farming on common lands, and backyard gardening.
6 Barbudans persisted largely through fishing during this time.
7 The land act was repealed by the national government in 2018.
8 Barbuda has an annual precipitation of 750–900 mm (30–35 inches). The IPCC scenario projections from 2007 indicate that temperatures will increase by 0.4 to 2.1°C by the 2060s, annual and seasonal rainfall will decrease, and sea level will increase globally at a rate of 1.7 mm/year (see McSweeney et al. 2016).
9 The salinity in Barbudan water even translates to language. Barbudans refer to two kinds of water: 'holy water' or 'good water' and 'pipe or well water'. Rainwater, which is collected from the roofs of houses and stored in cisterns or other containers, is 'good water' (Boger et al. 2014; Boger et al. 2016).

References

Antigua and Barbuda. 2008. "The Barbuda Land Act, 2007." *Official Gazette* 28 (5): 1–18. faolex.fao.org/docs/pdf/ant78070.pdf.
Antigua and Barbuda Census. 2011. Accessed September 1, 2018. http://apps.unep.org/redirect.php?file=/publications/pmtdocuments/-Antigua%20and%20Barbuda:%202011%20Population%20and%20Housing%20Census-2014Antigua%20and%20Barbuda_population%20census_2011.pdf.
———. 2012. "A Food and Nutrition Security Policy for Antigua and Barbuda." Accessed March 25, 2016. www.zerohungerchallengelac.org/ab/doc/Food Nutrition SecurityPolicyAG.pdf.
Antiguanewsroom.com. 2017. "Chinese Company Offers New Town For Barbuda." Accessed August 2, 2018. https://antiguanewsroom.com/news/featured/chinese-company-offers-new-town-for-barbuda/.
Beckford, C.L. and D.R. Campbell. 2013. *Domestic Food Production and Food Security in the Caribbean: Building Capacity and Strengthening Local Food Production Systems.* New York: Palgrave Macmillan.
Berleant-Schiller, R. Papers. *Various Dates: Documents at the National Anthropological Archives.* Washington, DC: Smithsonian Institution.
Berleant-Schiller, R. Papers. n.d. *Documents at the National Anthropological Archives.* Washington, DC: Smithsonian Institution.
Berleant-Schiller, R. 1977. "The Social and Economic Role of Cattle in Barbuda." *Geographical Review* 67: 299–309.
———. 1978. "The Failure of Agricultural Development in Post-emancipation Barbuda: A Study of Social and Economic Continuity in a West Indian Community." *Boletin de Estudios Latino Americanos y del Caribe* 25: 21–36.
———. 1991. "Hidden Places and Creole Forms: Naming the Barbudan Landscape." *Professional Geographer* 43: 92–101.
Berleant-Schiller, R., and L.M. Pulsipher. 1986. "Subsistence Cultivation in the Caribbean." *New West Indian Guide* 60 (1–2): 1–40.

Berleant-Schiller, R., and E. Shanklin, eds. 1983. "Grazing and Gardens in Barbuda." In *The Keeping of Animals: Adaptation and Social Relations in Livestock Producing Communities*. Totowa, NJ: Allanheld, Osmun, and Company Publishers. pp. 73–92

Boger, R., S. Perdikaris, A.E. Potter and J. Adams. 2016. "Sustainable Resilience in Barbuda: Learning from the Past and Developing Strategies for the Future." *The International Journal of Sustainability* 12 (4): 1–14.

Boger, R., S. Perdikaris, A.E. Potter, J. Mussington, R. Murphy, L. Thomas, C. Gore and D. Finch. 2014. "Water Resources and the Historic Wells of Barbuda: Tradition, Heritage, and Hope for a Sustainable Future." *Island Studies Journal* 9: 327–342.

Campbell, P. 2006. "Barbuda Sand Mining Proves Difficult to Manage." *Antigua Sun*, 24 October.

Cave, D. 2013. "As Cost of Importing Food Soars, Jamaica Turns to the Earth." *New York Times*. Accessed August 3, 2013. http://www.nytimes.com/2013/08/04/world/americas/as-cost-of-importing-food-soars-jamaica-turns-to-the-earth.html?pagewanted=2&nl=todaysheadlines&emc=edit_th_20130804&pagewanted=all&_r=0

Christie, M.E. 2008. *Kitchenspace: Women, Fiestas, and Everyday Life in Central Mexico*. Austin: University of Texas Press.

Coram, R. 1993. *Caribbean time bomb: the United States' complicity in the corruption of Antigua*. New York: William Morrow and Company.

Cornelius, T. 2007. *Golden Antiguans and Barbudans*. St. John's Antigua: Antigua and Barbuda Paradise, p. 108.

Crang, M., and I. Cook. 2007. *Doing Ethnographies*. London: Sage Publications.

de Albuquerque, K., & J.L. McElroy. 1995. "Antigua and Barbuda: A Legacy of Environmental Degradation, Policy Failure and Coastal Decline." In *Supplementary Paper No. 5*. Washington, DC: USAID, EPAT/MUCIA.

FAO. 2009. In McDonald, B.L. 2010:2. *Food Security*. Malden, MA: Polity.

FAO. 2012. http://faostat.fao.org/site/342/default.aspx.

Galhena, D.H., R. Freed and K.M. Maredia. 2013. "Home Gardens: A Promising Approach to Enhance Household Food Security and Wellbeing." *Agriculture & Food Security* 2: 8. http://www.agricultureandfoodsecurity.com/content/2/1/8.

Gordon, T. 2012. "Backyard Garden Initiative a Success." *Antigua Observer*, 10 May. http://www.antiguaobserver.com/backyard-garden-initiative-a-success/.

Gordon, T. 2013. "Stage Set for National Backyard Gardening Day." *Antigua Observer*, 30 May. https://antiguaobserver.com/stage-set-for-national-backyard-gardening-day/.

Harris, D.R. 1965. *Plants, Animals, and Man in the Outer Leeward Islands, West Indies; An Ecological Study of Antigua, Barbuda, and Anguilla*. Berkeley, CA: University of California Press.

Hart, M. 2010. "Nibbs' Sand Mining Comment Angers Marine Biologist." *The Daily Observer*, 27 August, 25–26.

Kimber, C.T. 1966. "Dooryard Gardens of Martinique." *Yearbook, Association of Pacific Coast Geographers* 28: 97–118.

Kimber, C.T. 1973. "Spatial Patterning in the Dooryard Gardens of Puerto Rico." *Geographical Review* 63 (1): 6–26. 10.2307/213234.

Kimber, C.T. 2004. "Gardens and Dwelling: People in Vernacular Gardens." *Geographical Review* 94 (3): 263–283. http://www.jstor.org/stable/30034274.

Lowenthal, D. and C. Clarke 1982. "Caribbean Small Island Sovereignty: Chimera or Convention? In *Problems of Caribbean Development*, edited by O. Franger. Munich, Germany: W FVerlag. pp. 223–242.

McSweeney, C., M. New and G. Lizcano. 2016. "UNDP Climate Change Country Profiles: Antigua and Barbuda." Accessed March 25. http://journals.ametsoc.org/doi/pdf/10.1175/ 2009BAMS2826.1.

McSweeney, C.F., R. G. Jones and B.B. Booth. 2012. "Selecting ensemble members to provide regional climate change information." *Journal of Climate 25* (20): pp. 7100–7121.

Oliver-Smith, A. 2005. "Communities after Catastrophe: Reconstructing the Material, Reconstructing the Social." In Stanley E. Hyland (ed.) *Community Building in the Twenty-First Century*. Santa Fe: School of American Research Advanced Seminar Series, pp. 45–70.

Perdikaris, S., R. Boger, and E. Ibrahimpašić. 2020a. "Seduction, Promises and Disneyficaiton of Barbuda post Irma. TRANSLOCAL." Culturas Contemporâneas Locais e Urbanas, n.°5 – Espaços (Des)Habitados | (Un)Inhabited Spaces, Funchal: UMa-CIERL/CMF/IA. translocal.cm-funchal.pt/2019/05/02/revista05/.

Perdikaris, S., R. Boger, E. Gonzalez, E. Ibrahimpašić and J.D. Adams. 2020b. "Disrupted Identities and Forced Nomads: A Post-Disaster Legacy of Neocolonialism in the Island of Barbuda, Lesser Antilles." *Island Studies Journal*. https://www.islandstudies.ca/sites/default/files/ISJPerdikarisBarbudaDisruptedIdentities.pdf.

Perdikaris, S., K. Hejtmanek, R. Boger, J. Adams, A.E. Potter and J. Mussington. 2013. "The Tools and Technologies of Transdisciplinary Climate Change Research and Community Empowerment in Barbuda." *Anthropology News*, February 2013.

Perdikaris, S., K. Hejtmanek, R. Boger, J. Adams, A.E. Potter and J. Mussington. 2014. "Connecting Transdisciplinary Scientists and Local Experts for Climate Change Research and Community-Based Adaptation in Barbuda, West Indies." *Anthropology News*, May 2014.

Potter, A.E. 2011. "Transnational Spaces and Communal Land Tenure in a Caribbean Place: 'Barbuda is for Barbudans'." PhD dissertation, Louisiana State University, Department of Geography and Anthropology.

Potter, A.E. 2015. "The Commons as a Tourist Commodity: Mapping Memories and Changing Sense of Place on the Island of Barbuda." In *Social Memory and Heritage Tourism Methodologies*, edited by S. Hanna, A.E. Potter, E.A. Modlin, P. Carter, D. Butler. London: Routledge. pp. 109–128.

Potter, A.E, S. Chenoweth and M. Day. 2017. "Antigua and Barbuda." *Landscapes and Landforms of the Lesser Antilles*, edited by C.D. Allen. Cham, Switzerland: Springer International Publishing, pp. 99–116.

Potter, A.E. and A. Sluyter. 2010. "Renegotiating Barbuda's Commons: Recent Changes in Barbudan Open-Range Cattle Herding." *Journal of Cultural Geography* 27 (2): 129–150.

Potter, A.E. and A. Sluyter. 2012. "Barbuda: A Caribbean Island in Transition." *Focus on Geography* 55 (4): 140–145.

Sauer, C.O. 1952. *Agricultural Origins and Dispersals*. New York: American Geographical Society.

Tandon, N. 2013. *Food Security, Small-scale Women Farmers and Climate Change in Caribbean SIDS*. International Policy Centre for Inclusive Growth One Pager. http://www.ipc-undp.org/pub/IPCOnePager220.pdf

Whitmore, T.M. and B.L. Turner II. 2001. *Cultivated Landscapes of Middle America on the Eve of Conquest*. Oxford: Oxford University Press.

Woods, M. 2012. "Rural Geography III: Rural Futures and the Future of Rural Geography." *Progress in Human Geography* 36 (1): 125–134.

6 Written with lightning

Filming Barbuda before the storm

Russell Leigh Sharman

The people in the old days, they used to call one another brother and sister.

Papa Joe sits on a rock just outside the wire fence that encloses his small yard and concrete block house. Tufts of white hair cling to his head and chin, his hands large and calloused and wrinkled with age. His eyes are a bit cloudy but still sharply focused. His granddaughter sits a few feet away, playing in the dust at his feet with the few baby chicks that scurry past. A few feet further still, a camera on a tripod. Papa Joe does his best to keep his gaze fixed on the woman asking questions, not the camera (see Figure 6.1).

'But they're not really brother and sister from the same mother and father', he continues. 'But they unite toward one another, so they say, they use the words brother and sister. And it was nice. Because what you have you share with me, and what I have I share with you'.

It's 2014, and my wife, Cheryl Harris Sharman, and I have recently landed on the small island of Barbuda to document as much of the traditional foodways of the inhabitants as we can in a few short weeks. After a choppy open-ocean crossing on a crowded pontoon boat, the only regular access to the island by sea, we're anxious to steady our sea legs and set to work. And as anthropologists and filmmakers, we know the quickest route to the thick description we're looking for is through a respected gatekeeper, a nexus in the local network of shared information. It helps if they've been around a while, lived a little. And in 2014, well into his 80s, Papa Joe had lived quite a bit.

He'd seen the island teeming with livestock, the soil tilled by nearly every inhabitant, and fishermen providing fresh fish and lobster from the calm waters of Codrington Lagoon and the open ocean beyond. The Barbuda of his youth, and much of his adulthood, was an island community of subsistence farmers, fishermen, and hunters. An island largely cut off from the rest of the Caribbean archipelago, but content to be so, providing for their own needs, and in some cases – such as peanuts and poultry – exporting their surplus to neighboring islands.

DOI: 10.4324/9781003347996-7

Figure 6.1 Papa Joe.

He'd also seen Barbuda pass into independence from Great Britain in 1981, along with sister island Antigua. He'd seen the island struggle in the shadow of that larger sister, the seat of power and post-colonial political and economic development. He'd seen the slow, inexorable slide away from subsistence foodways toward commodity capitalism until the only exports of value were sand and human labor, two of Barbuda's most precious, non-renewable resources. And he'd seen the passage of the Barbuda Land Act of 2007, a law that codified the tradition of communal land ownership for Barbudans and, at least on paper, protected their right to control economic development on the island.

But like many of the older residents of Barbuda, Papa Joe had mixed feelings about independence and misgivings about the enforceability of that 2007 legislation. Indeed, he made no attempt to hide a somewhat wistful longing for the benevolence of the Empire after years of perceived neglect and exploitation by home rule government.

'Now since the independence come now, it's worse', he says from his perch on that dusty road outside his home. His countenance shifts. The twinkle that was in his eye as he remembered days gone by dims. He stares hard at Cheryl just off camera.

> Because the people and them who in charge, who bring us independence, they promise to give them this and give them that and give the other. The whole thing is gimmick. People can tell you, 'Oh, I'll give you the world.' But they never live to see the world.

He pauses, a slight shake of his head. 'As the saying goes, "United, you stand, but divided you're bound to fall"'.

Papa Joe's lament would echo through every interview we filmed over the next few weeks. We would hear it in the pleas from working farmers and fishermen for more government investment and broader support for home-grown food security among their neighbors. We would hear it in the conflicted loyalties of local shopkeepers, witnessing first-hand that shift from subsistence practices to commodity capitalism, even as they profited from it themselves. And we would hear it in the convictions of educators, committed to raising a new generation of farmers and fishermen, a new generation of Barbudans who would call one another 'brother' and 'sister'.

Two years later, almost to the day, Papa Joe would pass away.

And a year after that, in September 2017, Hurricane Irma would devastate the island, wiping away all those farms and fisheries, all those shops and schools. Ninety percent of the built environment was destroyed. All 1,800 inhabitants evacuated to Antigua.

For the first time in perhaps thousands of years, Barbuda was a deserted island, empty of all human life.

But not for long. Within days, before government officials allowed a single Barbudan to return and survey the damage, a construction crew was on the island and clearing ground for a new international airport, a development pushed through Parliament without the consent of Barbudans in direct contradiction to the Barduda Land Act of 2007. Six months later, Prime Minister Gaston Browne introduced legislation to repeal the Land Act itself.

In the weeks and months that followed the storm, Barbudans would begin to make their way back, hoping to salvage what they could, rebuild, and start over. But with the likely repeal of the Land Act, the inevitable privatization of land, and the opportunistic, large-scale development projects imposed from outside by multinational corporations and the Antigua-based government, the hopes of farmers, fishermen, shopkeepers, and educators for a return to sustainable food security and self-sufficiency seemed to be slipping away. One only had to look to the massive earthmovers plowing through virgin forest land to make way for direct flights from the United States and Europe to visit resorts yet built to see it.

I was still sorting through 25 hours of footage when Irma hit Barbuda in 2017. Cheryl and I had edited a short documentary, *Sustainable Barbuda* (https://vimeo.com/199748662), about an aquaponics project started by anthropologist Sophia Perdikaris and John Mussington, a biologist and the local Secondary school principal. But there was still a trove of interviews to cull, edit, and make available. The goal was to put together a larger documentary about the traditional foodways of Barbudans as well as several stand-alone interviews of local farmers, fishermen, and hunters for research and archival purposes. Watching the reports of the devastation of the small island roll in, I knew whatever we had captured for those few weeks, just a few years before the storm, had suddenly become that much more important. The voices of Barbudans lamenting the loss of tradition and self-sufficiency and working toward a return to both were now re-contextualized. Irma had literally wiped the slate clean, but a larger political and economic storm was brewing in its wake. If there was any hope of weathering that second, larger storm, it would be in listening to the voices of those who had weathered the first, the shopkeepers, farmers and fishermen, the educators, and old-timers.

Old timers like Papa Joe who, back in 2014, summed up his feelings on the changes wrought on his island over his lifetime, then looked at Cheryl and I and said simply, 'You all wouldn't know. But I know. I grew up here'.

6.1 From subsistence to dependence

'I was born in Barbuda. And my parents, growing up, they ate everything they produced'. Miss Fancy sits in a back corner of her shop on an over-turned milk crate, framed by the half-empty shelves. A Winnie the Pooh beach hat perches on her head, clashing nicely with her nickname. No one uses her given name, Fancilla Frances. She's always been Miss Fancy (see Figure 6.2).

'We ate healthy, growing up we ate healthy', she continues.

> Because everything our parents grew here, that was food for us. And our meat came from the lagoon, which was fish, daily fish, that was it, from the lagoon. Daily. I never ate frozen fish, I never ate refrigerated fish. Every day it came from the lagoon.

Miss Fancy's shop sits just a few hundred yards from the airport. Not the controversial, still-under construction international airport, but the small landing strip for the semiregular, single-prop plane service from Antigua. Hers is one of a handful of shops on the island that sell foodstuffs imported

Figure 6.2 Miss Fancy.

from off-island – canned goods, frozen fish and chicken, some produce, and plenty of processed snack food and sweets. When we first arrived, we were told Miss Fancy was an important local contact, a prominent businesswoman and stalwart Barbudan. But at the time, she was visiting a daughter in North Carolina. Miss Fancy, like so many others, has seen more than one close family member leave the island in search of higher wages and a better future. Now many of them are bound up in transnational kin networks, spread thinly across the Caribbean, into the United States, and all the way to England.

Waiting for her to return, we spent some time on the dock, watching the daily ferry from Antigua – the same one we arrived on days earlier – and filming the arrival and unloading of the weekly cargo boat. The vessel was not much bigger than the small ferry, though it was heavily laden with boxes of inventory for the local shops, as well as building materials, appliances, bottled water, and dozens of propane canisters. This, apparently, was a new boat. A few weeks before we arrived, one of those propane canisters exploded on board the old cargo vessel, scuttling the boat and injuring many of the crew. Talk on the island was that the captain had been killed by sharks patrolling the wreckage. A grizzly story and evidence of the tenuous connection Barbuda maintains with her sister island and the rest of the Caribbean, tenuous but critical. With the cargo deliveries temporarily interrupted, supplies throughout the island grew worryingly low.

Days later, we're sitting in Miss Fancy's shop after her return from the United States, listening as she describes the quotidian struggle of any small business owner simply trying to keep goods on the shelves. It's late in the week, and supplies are running low. The weekly shipment is due that afternoon.

'They load up in Antigua and they leave about 12, 1 o'clock', she explains.

They should get here about 3:30, 4. But if they're late, it depends, throws them off a bit. The boat gets here, I go down. I have two trucks, because they offload the trucks, come up offload, return. And it's a long, drawn-out process.

She shifts a bit, hands on her knees. She seems tired just thinking about it.

Say the boat gets in at 5, I don't get to leave down there till about, maybe, sometimes 10 o'clock at night. And that is, that is really tiresome ... We offload the stuff from the boat down there, load it onto the truck, and truck it here to my store, offload it, get it into my store. That's one part of it. Then, tomorrow starts the process of gettin' rid of that stuff.

That 'stuff' includes a lot of staple goods, such as rice, flour, sugar, pasta, and other processed foods. But it also includes fresh produce, meat, and poultry, most of which is also imported from off island despite the fact that such things are produced locally by a handful of farmers. Miss Fancy explains that part of the problem is volume, that local farmers just can't match demand. At least not yet. But part of the problem is also the changing tastes of her customers.

> When I started out, with cold storage, I started out selling mainly [local] chicken, but then the taste came in for, like, American foods, like turkey and oxtail, and different stuff like that. [But] when I first started, we would sell the local chicken. Then the taste came in for the frozen chicken and we sort of phased out the local chicken.

Miss Fancy attributes some of these changes to those transnational kin networks extending far beyond the island. As more and more Barbudans leave the island to find work elsewhere, those who stay behind benefit from an increased flow of capital in the form of remittances. With more cash to spend in the local shops, there is less incentive to invest in the more traditional, labor-intensive subsistence practices. As she explains,

> Most of our parents and older siblings, they migrated out. They worked and sent home money for us, so we didn't need to work as hard, you know. So, we were able to buy other things. Mainly things that came in from outside. Imported.

And even though her shop depends on that shift in consumption, it worries her. 'I sell, weekly, about 52 cases of chicken', she says, dismayed. 'As opposed to the fishermen, you know, we're not eating as much fresh fish or fresh food anymore. Everything is imported. Most of it. Imported'.

Miss Fancy knows there are a handful of local producers trying hard to maintain those traditional practices as well as a government-supported farm designed to provide local produce. She does her best to support them all, selling as many locally produced goods as she can. 'But', she laments, 'we're still importing more of that than we are producing here. I don't know why. I really don't know why'.

A few blocks away, Ned Luke thinks he has an answer. He sits behind the counter of his own small shop, my camera catching him from across an aisle as Cheryl listens intently to his views on the subject. 'We were never like this before, pre-Independence', he explains. 'We were more self-sufficient, pre-1981'.

Ned is quite a bit younger than Miss Fancy, splitting his time between the shop he inherited from his father and the local secondary school where he works as a science teacher. Like many young Barbudans, he's spent time off the island, attending university in Cuba and working for a stint at a Sandal's resort on Antigua.

According to Ned, independence from England forced Barbuda into a new but perhaps equally paternalistic relationship with the government based in Antigua. A local Barbuda Council was given an annual budget and the power to dole out government jobs as a form of political patronage. Access to easy wage labor in exchange for votes, combined with the remittances already flowing in from relatives overseas, meant even less incentive to produce food for local subsistence, much less as a self-sustaining business. 'We didn't carry on the traditions of our forefathers. We didn't. And politics made things worse. Everybody can go work for the Council and get paid, you know. 'So why am I going out?'

Ned can remember the hard work of his father's generation that paid off in the form of fewer imports and even enough surplus to send out.

> My father used to plant cotton. My grandmother planted cotton. That building down there, they call it the ginnery, that's where we do the court, and the council meetings, that's what that building was used for. My father alone, in farming, he used to fill the cargo boats, him alone, with watermelons. You know, I mean, and they used to ship partly to Antigua, and then to St. Kitts, St. Barts, and St. Martin, and, you know, there was, there was some trade. You know, and the farmers here in Barbuda, they used to, they applied the traditional methods. They watched the moon, they looked at the cycle in the rainy season, and they do the mulching. You know, they used the older techniques. And they produced. We had an economy here.

Ned nods to several wooden bins just behind me and the camera, half-filled with a few items of fresh produce.

> Look at that rack. We have bananas and plantains, says product of Dominica. The garlic probably comes from China. The onions probably come from God knows where ... I would like to see everything on that rack be Barbudan. That's what we're hoping for.

As a teacher, Ned places a lot of that hope on the next generation. He does his best to instill a sense of pride in his students, a connection to the land and the sea, and their own past as a self-sustaining community. But Ned's hope is tempered by his own pessimism in light of the current political context, and like Miss Fancy, his role as a shopkeeper in feeding the consumer appetite for imported goods. 'You know, I mean, when we get down to the nitty gritty, we are teachers, we have failed this island, we have failed the community. I mean, we have failed.' He glances to the camera, then back at Cheryl, 'I mean, sounds funny to say, right, on the camera, but it's a reality'.

6.2 A return to the land

'One day I in my yard ...'

Eugene sits in a frayed and restrung lawn chair in the middle of his quiet, relatively isolated farm – his 'yard' – a mile or so outside the village of Codrington. He's surrounded by two or three acres of papaya trees, rows

Figure 6.3 Eugene.

of young corn plants, beans, and a few other vegetables and fruit trees he's lovingly planted over the years (see Figure 6.3).

'I in my yard one day', he continues in his lilting West Indian English, 'And I looked and a little boy passed my yard. And me hear the young boy say, "Wow, what a pretty yard". He said he'd love to get a place just like this'.

Eugene smiles, clearly pleased with the memory of the young boy who admired his farm. He speaks again, his voice so soft we have to lean in close to hear, and hope the microphone clipped to his faded t-shirt picks up the sound. 'You see, the young people, what they see you do, they will do … I wish more Barbudans come out and do this, so the younger ones can see, and go along'.

We've been at Eugene's farm for the better part of the day, hoping to get him to sit still long enough for an interview. We spent the first hour or so with a group of students from the United States touring the farm and learning a bit about Eugene's traditional farming practices. No pesticides, all organic fertilizer, and age-old techniques that maximize fresh water sources on an island plagued by scarce, relatively unpredictable rainfall. Getting usable footage surrounded by chattering undergraduates proved challenging, so we stayed back after they left, spent some time with him, and explained our goals for the project. He was eager to do what he could to encourage more Barbudans to turn back to farming. But as the camera started rolling, he froze, unable to speak. Not uncommon. Cameras have a way of unsettling people, especially those who've cultivated a life of quiet solitude, working the land, alone, undisturbed.

So we put in the time, passing the afternoon with Eugene, listening to his quiet patter about his life's work. By the time we started rolling again, he'd forgotten all about the camera, sitting in his favorite chair, under the shade of a slender papaya tree.

'I learned this from my parents, and my grandparents, and the rest Barbudans, the older ones, I learned this from them', he tells us. 'And I hope the younger ones do the same, take on the same pattern and do the same. I love to see a lot more come and do the same like we do here'.

Eugene goes on to detail some of the techniques he uses on the farm and their connection to the way farming has always been done on the island. His parents and grandparents were coal-makers as well, another deeply rooted tradition on the island. Using a few sticks and a tiny hole dug with his fingers, he demonstrates the process of baking coal in massive earthen pits. But the through-line of Eugene's quiet monologue is his desire for Barbudans to turn back to the land. For Eugene, the land is everything, and his tone turns rapturous as he tries to put this into words:

> I love this work. I love it. When I come, especially in the evening, and the morning, the early evening or in the morning, and you start to till the soil, turn over the soil, it make you feel good. Feel good. Really good. You know. I love to do that. And you look back by what I do, say, 'Wow, looking good, looking real good'.

Toward the end of our time together, that rapture shifts seamlessly into a prayer, taking Cheryl and I both by surprise: 'It's a blessed land we have here. All we just need to do, work the land ... Oh, Lord help me. Come on, young boys and young girls, get up and come over. Come out and farm'.

After a long moment of silence, his eyes scanning the tilled land around him, he turns to Cheryl, speaking of those boys and girls, 'I wish they could see and learn, you know. I wish that they could see and learn'.

Eugene is one of several Barbudans committed to maintaining the island's tradition of sustainable, subsistence agriculture. His farm, like several others in the hinterlands around Codrington, provides produce for the local shops, augmenting the imported goods from Antigua and beyond, as well as meat, poultry, and eggs to individuals throughout the island. But as Miss Fancy and Ned Luke point out, their collective efforts can rarely match the demands of local consumption, even on a small island like Barbuda.

One family hoping to change that is Shiraz and Anessa Hopkins. Along with their two young children, the Hopkins family runs a large farm raising cattle, goats, pigs, chicken, and Guinea fowl, along with various vegetable and fruit crops throughout the year. Both Shiraz and Anessa have dedicated themselves to learning as much as they can about agricultural science, while still holding onto traditional farming practices.

Their farm is an impressive operation if still a bit make-shift. Several pens were built from scrap lumber and sheets of tin, and an open pasture of hard-packed earth where cattle and goats roam freely. Fruit trees mingle with stands of scrub brush, all of it brittle under the bright sun. There hasn't been much rain lately. A worrisome development.

The day we meet them, we're with that same group of students, all of them eagerly peppering Shiraz with questions about his livestock. We've clipped a microphone to his t-shirt, tailing him with the camera, catching what we can as we walk and talk. 'I do this because of my grandfather', Shiraz explains.

> He had a passion for this … [But] most of his kids left for the States, Canada, England. My mom stayed back, so I was like, 'I'm not gonna let this go down the drain like that, I gotta do something to keep it alive.' And I just fell in love doing it. So that's why I'm trying to keep it going.

He shows us the pens holding various small livestock, and the new de-feathering machine he acquired to make chicken processing faster and more efficient. But it's the several heads of cattle that are clearly Shiraz's pride and joy. We catch him for a moment to talk about the operation, framing out the students, though their chatter provides a constant hum on the soundtrack.

'Anytime you go to the village and ask about my beef, you'll get a good answer about my beef because they like that, it's tender and everything like that', he says, beaming. But he knows, too, that quality doesn't always trump convenience. The cultural shift toward a cash economy and imported goods in the last few decades has forced producers like Shiraz and Anessa to actively promote a 'buy local' attitude. An odd problem to have on a tiny, isolated island, 35 miles across an open ocean from the rest of the Caribbean.

> The younger folks, they will easier go to the shop, they want to go quick, everything quick. But I want to change that. I want to bring it from right here, right out of Barbuda, from the earth right here, everything processed right here, so they'll get a much more natural and healthier food to eat … So I'm trying my best to try and let them come back to our local stuff.

Across a narrow track from the livestock is another fenced-in pasture, this one recently tilled and ready for seed. Anessa stands at the edge of the field of brown soil, describing their plans for the season. She's a bit more soft-spoken than her husband Shiraz, but no less enthusiastic about what they are trying to do for Barbuda.

> Most of the crops we plant are, like, tomatoes, green peppers, sweet peppers. We do watermelon, we do pumpkins, because we know those are in greater demand on the island. We choose crops that we know … the supermarkets will buy. And certain crops they prefer locally. Because our local sweet pepper tends to last longer than the ones we import … Ours have a longer shelf life.

Figure 6.4 The Hopkins family.

Several days later, we arrange a visit to their home, a modest concrete block house far enough from Codrington village to feel comfortably isolated, more connected to the land. Not surprisingly, they have a large fenced-in yard with some small livestock as well.

We set up the camera, framing Shriaz, Anessa, and their two children, Skylar and Shiresa (see Figure 6.4). All four of them work in the farm together, as a family. A fact that Shiraz is quick to point out:

> To have a family that farms together means a lot, because it comes from way back, from my grandfather. And just to see how he raised his family by doing farming, it's just so, so amazing. Now I have my family, and they're ready, they want to go, especially when it's time to harvest. They love that. When it's time to harvest. They're ready to go... From what my grandparents did, and my mother did, it just grew a part of me. So, I just like to do the same thing, and let my family see that there is hope in doing that. Because look at where it came from. There's a lot of hope in doing farming.

Shiraz and Anessa discuss what life was like growing up on the island when everyone kept a small 'ground' or farm, they used to cultivate basic provisions. As Anessa explains,

> Everyone back then was into farming, because that was the only way for them to provide. They didn't have a lot of goods coming in like we do now. So, they had to survive, and that was the means of survival.

They both go on to describe how that older generation developed an informal pattern of crop variety to both maximize expertise in one or two crops and satisfy everyone's subsistence needs through a barter system. The reciprocity established in the sharing of produce not only made sure everyone was well-fed, but it also confirmed and strengthened kin and non-kin networks on the small island, the 'brothers' and 'sisters' Papa Joe spoke of.

But as it had been made clear over and over in all of the interviews we had done to that point, times have changed on Barbuda. 'I have a cousin and he brings cargo from Antigua', explains Shiraz.

> And I go down there and I help him offload his boats. And I'm like, *every week* you have a big garbage bag full of anchovas, a big garbage bag full of cucumbers, a big garbage bag full of sweet peppers, boxes of tomatoes, *boxes* of tomatoes! And I'm like, 'Nah ... We have enough land here. We don't have a lot of people. We have enough land that we can cultivate to produce our own stuff.' So, I think we can and we should look into that ... We're supposed to be exporting, because we have the land to do it.

'Look at the other day when we had the cargo boat that sank', Anessa interjects, referencing the tragic accident when a propane canister exploded, sinking the vessel, and killing at least one crew member. 'We hardly had any produce for that weekend', she continues, Shriaz nodding in agreement.

> So, everything has to be brought in from Antigua or maybe Dominica, but if we had our own farmers, local farmers, then that wouldn't happen. So, in the event of anything happening like that, whether it's a natural disaster, like a hurricane, and we can't get any boats to come in, cargo boats to come in, what are going to do, you know, for a while? So, we need to start relying on ourselves instead of relying on other people for that.

Anessa makes a strong case for self-reliance, a return to a traditional intimacy with the land and what it can provide for the island's relatively small population, especially given their fragile connection to Antigua and the world beyond. But none of us knew then how prescient her reference to a natural disaster might be. A storm that could not only temporarily interrupt inter-island trade, but cut them off completely and potentially wipe out so much of the progress they've made in restoring that sustainable, subsistence practice.

But before all of that, back in 2014, many Barbudans were actually praying for rain. For generations, consistent rainfall had made traditional farming practices possible. But climate change has brought with it long stretches of drought, punctuated by increasingly severe hurricane seasons. To make matters worse, the good system employed by many on the island, including the Hopkins family, suffers from increasing salinization as rising ocean levels begin to penetrate groundwater. And that's if you have the money to fuel your gas-powered pumps.

Back on his farm, Shiraz inspects the brittle leaves of his fruit trees. 'Right now I'm just praying for the rain', he confesses.

> We need some rain right now. I have a well and I have a pump and everything. But it's too expensive for me to pump water like everyday, to water these areas. So, I've gotta just pray for rain and hope for the rain to fall.

There's a nervous chuckle in his voice, a worry that's hard to conceal. He pauses a moment and looks off to the dry, dusty earth around us. Then, as if remembering his role as self-appointed ambassador for a return to farming, he looks back to the camera and brightens, 'But I love doing it, no problem. I love doing this and don't think I'm gonna stop. I'm not gonna let my grandfather's work go down. I'm gonna keep it up'.

6.3 Coming home to roost

'The whole of this area is round about, um, almost thirty acres all together', explains Frances Beazer with a wave of her arm. She's moving across an open pasture, heading toward a large concrete block and corrugated tin building that dominates the open space. The muted cacophony of hundreds of clucking chickens rumbles behind the tin walls in the distance. 'The chicken farm occupies a very small area', she continues. 'The whole area belongs to the farmer's cooperative. But we are the only ones that are active at the moment'.

Frances moves slowly, but with purpose; a baseball cap fits snugly over her short-cropped white hair, shielding at least some of the afternoon sun. Well into her sixties, raising chickens is Ms. Beazer's second career. Like so many others, she left the island years earlier, eventually landing in England, where she spent several decades as a nurse and then a social worker. But unlike many of those who left, she eventually returned home to Barbuda, choosing to spend her retirement years building up a sizeable egg production facility.

We stay a few paces behind her as she walks, the camera following her on her daily rounds. 'We created this chicken farm three years ago', she tells us.

> The purpose for doing so is to let young people see that chickens don't always have to come from abroad, that we could create chickens here. Chickens can be born, eggs can be hatched, and we can supply the island with as many eggs as they need. However, young people don't appear, at this moment, to be very much interested. But we are hoping that eventually they'll see the value of what we're doing and will come along and join us.

Frances has not let the apparent lack of interest among the Barbudan youth to slow her down. When her hens are at peak laying season, her farm

Figure 6.5 Frances Beazer.

produces as many as 20 dozen eggs every day. Nearly 250 eggs that she and her husband collect, clean, and package for the local island market, day in and day out. Miss Fancy is one of her best customers. Still, her primary motivation was and is to encourage others to re-engage with the land. Like Shiraz and Anessa, she hopes she is leading by example, but she knows there are many obstacles to overcome.

'It's expensive living on an island', she admits, echoing what so many others have told us over the past several days. 'Especially an island like Barbuda. Quite expensive. Because you have to pay for everything that comes in'. We're standing in a shipping container a few yards from the hen house where Frances carefully washes each egg in a bleach and detergent solution, then packages them by the dozen (see Figure 6.5).

> When I was a child, we used to grow enough vegetables and provisions to feed ourselves. But things have changed. The soil doesn't yield as much as it used to... And, as people become more westernized, they think it's old fashioned to work the land, to do farming.

Frances dunks another egg into the large yellow plastic washbasin, wipes it gently with a cloth, and then sets it to dry on an overturned shipping barrel. All of it with practiced grace, the smooth motion of an action repeated too many times to count. 'They prefer to go and work in offices', she continues, 'where you can dress up in suits, and ladies wear high heels and nice minidresses and show their legs and all that sort of thing. So, farming in Barbuda is on the decline. We're trying to revive it, but it's, it's a very lengthy process'.

She goes on to describe some of the particular challenges of farming in Barbuda. The difficulty of tilling the soil with limited equipment, the reliance on traditional techniques that require more labor than modern, industrial farming practices, and perhaps most challenging, the recent lack of rain. She's quick to point to climate change as a likely culprit, but she also acknowledges that Barbudans have not always been good stewards of the land; clear-cutting trees without a plan for reforestation, for example. Or the rampant population of feral donkeys and other grazing livestock that eat through the protective ground cover.

But Frances is just as quick to praise her home island and her people. 'Barbuda, in my opinion, is a unique place', she tells us.

> I don't think there's any other place like it. Even the people. I don't think there's any other people like Barbudans. We're knowledgeable. I'm not saying we're educated, but we're knowledgeable. And when we hold an opinion, it's very difficult to shake that opinion out of us. Everything is fresh. We get fresh air. We have fresh food, when we plant it and cultivate it. Fresh fish, from the sea. The land is free. We got our land act passed in 2007, in the Houses of Parliament in Antigua, and the land belongs to the people. I, as a native born, I am entitled to a plot of land for housing, a plot for grazing, and a plot for cultivation. Where else could I get that?

Frances smiles, a twinkle in her eye as she reflects on the central role the land has played in Barbudan identity, and the critical importance of the Barbuda Land Act of 2007 in protecting residents' connection to that land. Like Anessa Hopkins's oblique reference to the hypothetical dangers brought by unpredictable Atlantic storms, Frances's reference to the Land Act would prove prescient. In the aftermath of the storm, with the Land Act under attack by Antiguan politicians, her question – 'Where else could I get that?' – will loom large.

In the meantime, however, Frances is more concerned with her primary goal of attracting young people to working the land and reconnecting with sustainable foodways.

> Going back to when I was a child, we had to learn how to grow things. We were introduced to planting things in the soil, and watched them grow. Watered them and so on so forth. And I think they should go back to that. And teach children... Let the children see that we can grow what we eat... Because some children, they don't realize that certain things are grown. They think you just buy them like that. And I think this is what the schools should do.

She pauses a moment, focused on the last few eggs she has to clean. She's a dreamer, but she's also a realist. 'And all of them, naturally, is not going to become farmers', she admits. Then, hopeful, determined, she adds, 'But some will'.

The days' harvest of eggs cleaned and packaged, Frances stands at the open doorway of the shipping container, surveying the ruddy red chickens scurrying here and there as they enjoy a few more minutes in the open air. Soon her husband will rap on the metal wall of the container, and their flock will stream back toward the hen house, as well trained as they are well cared for.

She takes this moment to give us a little background on the farmer's cooperative that owns the land we stand on, of which she is a founding member.

> In 2005, we started the cooperative, on paper. A group of us got together because we were very much interested in agriculture and want to see agriculture revive in Barbuda. Most of us were middle aged people, so to speak, middle age to elderly. And each one had a desire to do one specific thing. Well, I always wanted to produce eggs. Some people wanted to produce the chickens for the dinner plate. Some wanted to produce vegetables. Some wanted to produce roaming animals, small stock, like pigs, and, uh, Barbuda is famous for rearing goats and sheep and things like that. So, people had an idea what they would do. But of course, like everything else, people change their minds.

She looks around the empty field, cleared and ready for crops if anyone cared to sow them.

> And, um, as you see, I'm the only one at the moment doing something. Other people are still in the process. They say they're going to do this, and they're going to do that, but we have to wait and see.

Somehow, she seems undaunted by the lack of involvement of the other members of the cooperative. Or the apparent lack of enthusiasm of Barbudan youth. For now.

> My dream is that we, Barbuda become productive; that we're able to produce the food that we eat, all locally right here. And I think we can. Yes, I think we can do it. With a bit of willpower. And the rain from above. I think we can.

6.4 Captain Speedy

'Well fishing in Barbuda to me is … Well, it's a way of life.'

Arthur 'Speedy' Walter sits at a wooden table in the open air by the small pier that serves Codrington village. Behind him is a small, protected marina. A few vessels bob in the dark green water. Some are large enough for the open ocean. Some are meant for fishing the calm waters of the lagoon (see Figure 6.6).

'I've known fishing all my life', he continues. 'My family were fishing people, boat builders and fishermen'.

Figure 6.6 Speedy.

Speedy is in his early 40s, broad-shouldered, tall, a body built for the hard labor of longline fishing, trapping, and conch diving. A frayed cap covers his nearly-shaven head. A pair of sunglasses is on the table in front of him. When he talks, his hands are always in motion, his arms swinging in wide arcs to emphasize a point.

'I used to move around by boat a lot', he explains, a bit wistful. But maybe also a hint of regret that it took him as long as it did to settle down.

> Now I'm sitting here in Barbuda, I'm married now, I have a daughter, and I'm building a business. I was never grounded. I got the name Speedy because I'm always on the go. But now it's a whole different picture. Barbuda is a beautiful place. It's quiet. You can think here. Get on with living. It's very important for young people who get carried away with the fads and the new lifestyles, the crazy outside [world]. You can find peace of mind here. You can find a good living here. You can raise your children here. A healthy place. A natural place.

Speedy is part of a community of fishermen working the waters of the lagoon and the Caribbean beyond. For most of our time on Barbuda, we were stymied in our attempts to arrange an interview with him or any of the other Barbudans who make their living on the water. Mostly because their schedules never leant themselves too much time on land talking in front of a camera. Speedy proved worth the wait. His passion for his work was matched only by his passion for preserving and protecting the fisheries and passing his knowledge down to the next generation. 'I think we can teach the children to take care of what they have', he tells us. 'Which is the important thing. Take care, not mash it up'.

Speedy favors vertical longline fishing for snapper and squid. Mostly because he can do it on his own, without a crew. But he also works lobster traps along the reef and deep-water conch diving. Both of which require at least one other hand on deck, if not an entire crew. Like most fishermen, his day begins early, at 3 am or 4 am in the morning. He's on the water by 5 am, before the sun rises. And if he's diving for conch, he's gone until sunset.

'Conch is a lot of work', he explains. 'Taking it out of the shell, collecting them, cleaning them afterwards. It's a full days work, conch diving'. And as he describes the difficulties in diving and processing conch, he shifts effortlessly into the importance of conservation. A theme he returns to over and over.

> I remember some years ago they thought conch would never be done. But it's at a point now where it's a protected thing. And it has to be because the conch beds only have so much yield and it takes so many years for a conch to develop to an adult, to maturity.

For Speedy, conservation is not a burden to be managed but a critical part of his vocation. And he recognizes that passing that ethos on to other fishermen is a critical part of maintaining productive fisheries around the island. His evangelism on the topic rivals that of the farmers in their attempts to turn people back to the land. In fact, Speedy himself makes a direct connection between fishing and farming:

> It's very much like farming. A farmer will never kill his prized bull, that's in the taking of fish and lobster and conch. He'll never kill his prized bull, and he'll never kill a cow that's in calf or he doesn't kill animals with growth potential. So, what you have there then, you create a balance between those three points. And when you really start to look at it, you begin to realize that the ocean is just teeming, life everywhere.

But he also recognizes that it can be a hard argument to make with men and women who depend on productive fishing to provide for their families. He sees the value in tightly controlled seasons for lobster and conch to allow breeding and maturity. But he also knows that a lot of his colleagues have struggled to make ends meet, especially lately. Foreign fishing vessels have started to troll dangerously close to Barbuda's shores, capable of plundering what stock remains near enough for small boats to reach. Worse, some foreign fishermen have taken to using bleach to kill the reefs and stun the fish, making them easier to catch. And making it more and more difficult to make a living for local fishermen. 'It's a balance that has to be struck. We're living by it, but we also have to protect it at the same time. And it's not an easy thing when you're at a certain state of desperation financially.'

Like most of the Barbudans we've talked to, Speedy sees hope in the next generation, the younger folks who will have to choose between wage labor off the island, or making a living off the land and sea like the generations before them.

> I think that if you open that education to younger people in the school level coming up, people will grow with it. They will understand it better. And that way they will grow with a love and an understanding for the ocean that will replace indiscriminate taking and have the natural conservation idea. Conserve. Protect. Take care of. And it will be a beautiful place, for all of us.

But he also understands the call of the wider world. He spent enough time away to know its attraction. But as with conservation, he's convinced it's all about balance. 'For the young people, go out and get experience', he encourages,

> but come back with it and use it at home. Don't neglect your home. Try and help the younger ones that are coming up to give them the understanding. Because all of us can appreciate that it is here and here that it has to start. And I know the young people in school today, I think they're learning that too. We can help each other, the younger generation by showing them a better way. There is a better way for all involved. For the stakeholders and the environment.

6.5 After the storm

'Things really not what they used to be, and it's getting worse'.

Papa Joe is still on his rock; the sun starting to dip toward the horizon, the granddaughter has long since run after the baby chicks that scurried into his yard. We are nearing the end of our time together, his narrative bending back more and more often to that central complaint about how much things have changed for the worse since the halcyon days of his youth. It's a familiar and perhaps universal complaint of the elderly, lamenting the passage of time. But Papa Joe is more specific. For him, the root cause is discontinuity and a lack of history.

'Because the young people today think what is going on now, that is what going on in time gone by. But it wasn't so ...', he explains. 'They doesn't know anything about it, you understand. Because nothing there written in history or anything that say, ok this thing happened so and so a time'.

Like the others we've talked to, Papa Joe feels some responsibility to pass on that history. But it's tinged with a pessimism that also seems to come with age. 'So, now, you try to tell them about things that happen in time gone by, you know what they say?' He asks, looking at Cheryl with a rueful smile. 'They say, "Oh, you old time-ish"'. And he falls into a raspy chuckle. It takes him a moment to recover from his own joke. 'You understand me I say? Because they don't believe you in what you're trying to tell them. They don't know what you tell them is what you passed through'.

Soon he's back to his other favored topic, politics. The mirth is gone. The pessimism is weighing on him as he hunches over on his perch. 'And the worst thing that is happening now, is family against family, is politicians come in and divide us all up, you know'. He shakes his head, staring at the dry, dusty ground at his feet, 'So, I don't know what is going to happen, but it's going to be terrible'.

Like so many of the men and women we interviewed during our short time on the island in 2014, Papa Joe's comments seem oddly prescient now in the context of hurricane Irma's devastation. Mr. Mussington, a biologist and local Secondary school principal, showed us the importance of mangroves in protecting the island from dangerous storms, and his own multi-year project to introduce agriculture, egg farming, and aquaponics to his students. To Alcon Ned and Shaville Charles, both hunters, among many other things, upholding the tradition of protecting and sustainably culling the island's unique population of wild deer, hogs, and land turtles. To the many other farmers, shopkeepers, food vendors, educators, and fishermen we spoke to. All of them seemed to call for a reconnection to the past, a shoring up of community and solidarity in the face of an uncertain future, bracing for changes to come. Of course, none of them could have known how radical those changes would be, first from Irma, then from the political and economic forces that would sweep in to take advantage of Barbuda's weakened state.

Papa Joe didn't live to see Irma lay waste to his island, and perhaps that is for the best. But for Miss Fancy, the Hopkins family and all the rest, it has been a test of their resolve to make Barbuda the sustainable, self-sufficient island they always hoped it could be. A year later, Miss Fancy's store had not re-opened. She was still cleaning out the wreckage, repainting and repairing the damage. The Hopkins family were some of the first to return to the island, quickly repairing and rebuilding their farm. Eugene was back as well, living in his 'yard' and carefully restoring what was lost. Frances Beazer returned to find that many of her beloved chickens had died in the storm, but the few that were left came running to greet her. The hen house was repaired and resupplied with the help of aid organizations. Speedy's new business, a seafood restaurant, was completely destroyed. He had plans to rebuild, but in that first year after the storm, he spent most of his free time running boats back and forth to Antigua, helping others with supplies for their own recovery efforts. Some we spoke to, however, did not return at all.

In the storm's aftermath, with Papa Joe's lament and the hopes and dreams of all those we interviewed still ringing in our ears, we went back to the footage, hoping to preserve the voices of Barbudans before the storm in a series of standalone films: Papa Joe (https://vimeo.com/478991277), Eugene (https://vimeo.com/478968547), Frances Beazer (https://vimeo.com

/478969681), Shiraz and Anessa Hopkins (https://vimeo.com/478986711), Speedy (https://vimeo.com/478993695), along with Blaine 'Righteous' Frank another small farmer (https://vimeo.com/478992606), and hunters Alcon Ned and Shaville Charles (https://vimeo.com/478989337). All of them culminating in the documentary *Before the Storm: Subsistence and Survival on the Small Island of Barbuda* (https://vimeo.com/478959977).

And in the process of transcribing and editing hours of footage, listening to all of those voices again and again, one thing became abundantly clear: the consistent call for a return to the land and sea, for self-sufficiency and sustainability, was not, in the end, about food at all. It was about community, identity, and solidarity. When Barbudans turned away from the land and looked to the ocean, not for what fishermen could provide but for the boats carrying goods from Antigua, they turned away from a system of mutual support and reciprocity that meant far more than food on the table. They turned away from the fictive kin network of 'brothers' and 'sisters' that Papa Joe described: 'What you have you share with me, and what I have I share with you'. For generations of Barbudans, community and identity were built on a foundation of interdependence. But without traditional practices like farming, hunting, and fishing that encourage those crucial systems of reciprocal exchange and interdependence, words like 'community' and 'identity' become abstractions, powerless against the radical infrastructural and political changes that threaten Barbuda's autonomy. That threatens its very future.

Solidarity, then, among Barbudans is more important than ever. And in the end, self-sufficiency and sustainability may be as powerful a political act as it is a practical concern.

The final documentary, *Before the Storm*, preserves these voices from a time that is now, in Papa Joe's words, 'A time gone by'. It captures what is lost, but hopefully, not lost forever. As the last title card in the film reads:

> The future of the island depends upon listening to these voices, the voices of Barbudans, respecting their right to self-determination, and protecting their connection to the land that has sustained them for generations.

7 Disaster capitalism

Who has a right to control their future?

Emira Ibrahimpašić, Sophia Perdikaris,
and Rebecca Boger

7.1 Introduction

Every disaster, whether it be war or a natural event like a hurricane or flood, presents a set of challenges but also opportunities. For the people in power and those with the economic means to do so, disasters present opportunities to solidify that power and further fill their coffers. The vulnerability, trauma, and confusion that follow the aftermath of these tragic events create ideal conditions for disaster capitalism to flourish. Oftentimes within days of a disaster, we see bulldozers working away at the disasterscape, even before any aid or relief has arrived (Boger and Perdikaris 2019; Gould and Lewis 2018). This moment of transition, a space in which local inhabitants (both human and animal) are simply trying to figure out how to orient themselves, is exploited for one simple goal – the endless pursuit of profit.

The dominant narrative that defines the current global economic order is rooted in the liberal economic theory first championed by Adam Smith. Smith proposed that each nation can achieve wealth through the efficient division of labor and use of resources (Coats 1979). Resources, specifically scarce resources, directly affect the creation of wealth, and therefore those who control resources control wealth and therefore power (Coats 1979). In the case of Barbuda, the scarce resources come in the form of its land, which was left relatively untouched by development. As a result, Barbuda has attracted interest from the central government as a space where profit can be made, irrespective of cultural norms, environmental destruction, and against international treaties and agreements already in place. Barbuda's landscape and seascape with extensive forests, beaches, and coral reefs have and continue to be turned into a commodity for sale to the highest bidder. With relatively few unspoiled places left around the globe, wealthy foreigners seek to capitalize on this (Perdikaris et al. 2021a, 2021b) with complete disregard for long-term consequences and the demise of local lifestyles and customs. To achieve their goals of a quick profit, the Antiguan government has used the disasterscape of post-Hurricane Irma as grounds to dissolve centuries-old understandings of land/people relationships that were codified in the 2007 Barbuda Land Act. The Land Act protected Barbuda and

DOI: 10.4324/9781003347996-8

its inhabitants and allowed them to maintain their ways of life relatively undisturbed by the foreign infringement (*The Guardian* 2020) for nearly 400 years until it was repealed in 2018.

The disasterscape like the one in post-Irma Barbuda offers a 'fertile ground for implanting particular neoliberal policies because of an acute temporary crisis that demands quick action and usually an infusion of emergency assistance' (Stonich 2008, 22). In Barbuda, just like in many other places around the world, local governments seized 'the opportunity to shut down participation, implementing decisions without debate' (Stonich 2008, 22). The repeal of the 2007 Land Act has allowed the Antiguan government to push through with development projects that will permanently transform Barbuda's landscape, environment, and ways of life (see Figure 7.1). In addition to devastating effects on people, their history, and their heritage, disaster capitalism poses a challenge to sustainability and long-term human occupation. Sustainability refers to the 'development that meets the needs of the present without compromising the ability of future generations to meet their own needs' (Purvis et al. 2019). As we will show, the current neoliberal policies occurring in Barbuda and elsewhere are not able to achieve the goals of sustainability and will permanently transform the island and the lives of its inhabitants.

The loss of long-existing communal land ownership harms Barbuda's most vulnerable, and will have long-standing effects on their livelihood, as well as their tangible and intangible cultural heritage. What this case also highlights is the disconnect between state and people, and between government promises and practice. The failure of the Antiguan government to honor international and domestic agreements further highlights the lack of respect for local needs and wants. Ultimately, this is not only damaging to the local people but puts the validity of the Antiguan government in question on a global scale.

In this work, we examine the effects of disaster capitalism broadly and how it's been progressing in Barbuda. Further, we analyze the effects of disaster capitalism on Barbuda's environment and ecology, as well as its people and their tangible and intangible heritage. Through this discussion, we consider problems associated with present-day development with special attention to increased vulnerability through disaster capitalism and neoliberal economic policies.

7.2 Disaster capitalism and Barbuda

Hurricane Irma made its landfall over Barbuda on September 6, 2017. Within a couple of days, the Prime Minister visited Barbuda and ordered an evacuation. What was initially seen as an act of kindness quickly revealed itself to be the opposite. A few days after the mass evacuation of Barbudans, foreign workers with heavy machinery were on the island working around

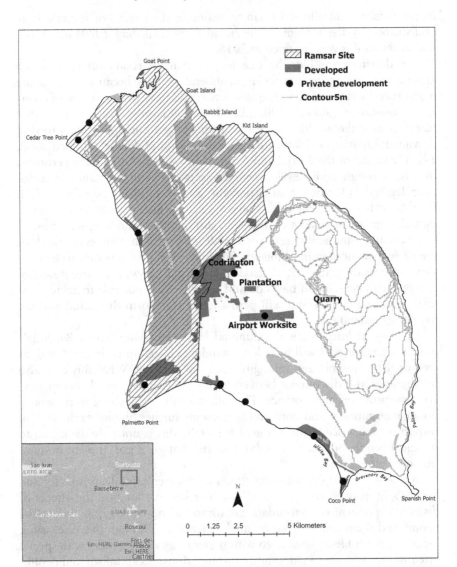

Figure 7.1 Map of Barbuda showing the Ramsar site, airport excavation, and recent or proposed tourism developments mainly financed by John Paul Dejoria. Two sites are within the Ramsar protected wetland area.

the clock. Instead of focusing on island recovery, they began clearing the forest for the construction of a new and controversial international airport (see Figure 7.1) (Boger and Perdikaris 2019; Perdikaris et al. 2021a, 2021b). By October 22, 2017, less than six weeks post-Irma, the footprint of the new airport was already a third of its current size. While native Barbudans who attempted to return were scrutinized by the government and actively prevented from returning to their homes, hundreds of non-Barbudans, including scientists, reporters, military, and construction workers, came and went at their pleasure. The Prime Minister kept Barbudans out so that they would not bring disease to Antigua as a result of flooding and dead animal carcasses that still littered the island (Chappell 2017). While Barbudans were kept out, the Prime Minister proceeded with his plans to develop the island. All of this was done without proper approval or discussion by the Barbuda Council, the local governing body, and other local stakeholders; non-Barbudans proceeded with their agenda of constructing high-end luxury tourist infrastructure.

The actions by the Antiguan government at the time were illegal because they were in direct contradiction to the 2007 Barbuda Land Act that was still in effect at the time. While initially there may have been some support for the airport, the fact remains that the people of Barbuda had not made their choice before the construction began. Four years after Irma, it is apparent that the events that took place immediately after the hurricane were premeditated. The speed at which the airport construction began post-Irma indicates that the funds were already planned and available. The project proceeded without proper environmental assessment failing to meet the proper Federal Aviation Authority (FAA) requirements. In spite of the lack of structural integrity, the airport is prominently featured in all proposals for development on Barbuda as an integral island infrastructure and available for use after new roads are cut that will lead straight to the new site locations (Brosnan, 2020). These events confirm that the Prime Minister was in conversation with investors long before the hurricane and used Irma to accelerate his agenda.

The flurry of activities in Barbuda immediately after Hurricane Irma is what Naomi Klein refers to as 'shock doctrine', the 'deliberate exploitation of states of emergency to push through a radical pro-corporate agenda' (Klein 2018, 45). The events that have taken place in Barbuda mirror those of post-Hurricane Maria in Puerto Rico (Klein 2018; Boger et al. 2019) where the powerful elite used the disasterscape as an excuse to suspend democratic decision-making. The other part of shock doctrine is that it is fast 'pushing a flurry of radical changes through so quickly it's virtually impossible to keep up' (Klein 2018, 45). In Barbuda, just like in many other parts of the world, the shock doctrine is best described as 'orchestrated raids on the public sphere in the wake of catastrophic events' that lead to 'the treatment of disasters as exciting market opportunities' (2018, 6). Klein provided a definition for a phenomenon that hundreds of other scholars have observed across the globe and sit at the core of the expansion of capitalism and neocolonial economic expansion practices that define post-World War II global economic order.

The shock doctrine is part of a larger set of policies and agendas that are part of the global capitalist DNA. Sometimes, the shock treatments occur as a result of a natural disaster like in the case of Puerto Rico, and sometimes they occur as a result of profound social and political changes. In the early 1990s, we saw a similar shock doctrine across the former Soviet Block. The newly liberated states found themselves in a flurry of activity that resulted in mass-scale privatization of once state-owned industries; this significantly impacted peoples' economic livelihood and left many without any means to support themselves (Manzetti 2009). Shock doctrine is an important tool of another set of neoliberal capitalist policies, most notably disaster capitalism. Disaster capitalism examines different ways in which 'economic shock treatments' aimed at the pursuit of free-market economic revolution have caused people to lose land, homes, and livelihoods to corporate makeovers (Manzetti 2009).

Schuller and Maldonado define disaster capitalism as '[n]ational and transnational governmental institutions' instrumental use of catastrophe (both so-called "natural" and human-mediated disasters, including post-conflict situations) to promote and empower a range of private, neoliberal capitalist interests' (2016, 62). What makes disaster capitalism so power-ful and its effects so disastrous is 'the instrumentality of catastrophes for advancing the political, ideological, and economic interests of transnational capitalist elite groups' (Schuller and Maldonado 2016, 63).

Those in power see these moments of confusion and instability as oppor-tunities for profit, and they have successfully been able to exploit country after country leading to further divisions between the rich and the poor and creating new elites whose only goal is the pursuit of profit. In Barbuda, the privileges of exploiting the island were given to

> [a] group of investors, including John Paul Dejoria, the billionaire entre-preneur behind Paul Mitchell hair products, have been given a 99-year lease to create hundreds of deluxe private homes and a golf course for the scheme named Peace, Love and Happiness (PLH).
>
> (*The Guardian* 2020)

Disaster capitalism, in addition to implementing one or more of the parts of the Washington Consensus, including privatization, liberalization, and fiscal austerity, ultimately leads to 'uneven gains of rapid privatization' (Gunewardena and Schuller 2008, 7–8). The reality of the events in Barbuda highlights that 'disasters are not primarily natural events but political events' (Gunewardena and Schuller 2008, 17). The actions taken by the Antiguan government further produced hazards and exacerbated vulnerability by destroying wetlands and undisturbed animal sanctuaries, thereby 'amplify-ing the storm's destructive efforts' (Gunewardena and Schuller 2008, 17).

What unites the experiences of Barbudans with those of people around the globe who have faced major (and sometimes multiple) climate change-induced events is the disregard for local people's opinions and voices; this has devastating effects on the ability of the local community to survive, such as when Hurricane Mitch made landfall in 1998 on Honduras. In the after-math of Hurricane Mitch, the government of Honduras used the disaster to enact various economic and political changes that are clearly identified as 'disaster capitalism' (Stonich 2008). Similarly, the Antiguan government actively exploited the confusion in the post-Irma disasterscape. The entire population of 1,800 Barbudans were forcibly evacuated to Antigua and kept away from their homes. They were excluded from decision-making while the central government proceeded with its development agenda.

Perhaps what is most dangerous in these situations is the suspension of democratic values and principles that so many countries have paid a high price to achieve. The events in Puerto Rico, Honduras, and Barbuda demonstrate the fragility of democratic institutions worldwide. The authoritarian tendencies of the leadership in these countries further exemplify just how many governments resort to entering into monologues rather than dialogues. This type of behavior leaves local populations without a voice and presence at the negotiation table where the decisions about their future are made. Leaving individuals and communities out of the dialogue is dangerous because it allows powerful corporations to make decisions about peoples' future (Global Legal Action Network 2020).

In addition to the destruction of traditional lifeways, disaster capitalism leads to further division between the rich and the poor (Birau and Doaga 2019; Bourne 2009; Edward and Sumner 2018; Yee 2018). The building of airports, hotels, and other tourist attractions pushes the local populations into the lives of servitude (Perdikaris et al. 2021b; Perdikaris and Hejtmanek 2020). This type of economic subsistence bears resemblance to economic practices and lives during the colonial period and slavery. The promise of economic growth through the development and implementation of the western free-market order has in recent years been called to question as the gap between the rich and poor continues to grow (Birau and Doaga 2019; Bourne 2009; Edward and Sumner 2018). In those 'developing' countries with a high percentage of economic growth, we've seen a continued widening gap between the rich and poor (Bader et al. 2017, 2068).

Much like the rest of the world, the Caribbean has seen some economic growth over the last several decades. Despite these changes, poverty remains a major problem. The Caribbean Development Bank admits that despite the economic growth

> [E]conomic volatility causes fluctuations in employment and incomes, with particularly strong influence on the employment and incomes of lower skilled workers. The poverty effects are magnified because poor people have weaker and less effective mechanisms for coping with loss of employment and income.
>
> (Bourne 2009, 24)

These weaker mechanisms are a result of the uneven economic growth between the rich and the poor, uneven distribution of the country's income, lack of social net mechanisms, and little investment in the education and training of the workforce. Moreover, much of the economic growth tends to rely on the growth of one single industry, which makes the countries vulnerable in the long term. In the Caribbean, the heavy reliance on tourism makes the countries more vulnerable and unable to cope with large-scale economic fluctuations (Perdikaris and Hejtmanek 2020; Perdikaris et al. 2021b). Without a doubt, the events that have unfolded in Barbuda resemble what occurred in post-Hurricane Maria Puerto Rico where disaster capitalism was used to create a safe, albeit manufactured, playground for millionaires and other elites (Klein 2018).

Antiguan government, seduced by the quick profit promised by the wealthy outsiders, ignores mounting evidence that increased reliance on tourism will lead to increased dependency, gentrification, and a growing divide between haves and have-nots. One of the best examples of the failures associated with the implementation of the tourist industry as the primary means of subsistence is found in Honduras (Stonich 2008). In post-Hurricane Mitch, the Honduran government embarked on larger-scale tourist development projects that resulted in relegating

> poor Afro-Caribbean islanders to low-status, low-paid, temporary jobs; redacted access for local people to natural terrestrial and marine resources on which they depend for their livelihoods; escalating prices for food, manufactured goods, and housing; land speculation and spiraling land costs; increased outside ownership of local resources; and deterioration of the biophysical environment--especially declines in the quality and quantity of fresh and seawater, deforestation, and habitat destruction.
>
> (Stonich 2008, 58–59)

This is what will likely happen to Barbuda.

It appears as if the Antiguan government has not done due diligence by choosing to focus on luxury tourism as a single source of revenue (Perdikaris and Hejtmanek 2020). It is as if the world has learned little from previous failures of this type of tourism and its effect on local cultures. Instead, the central government in Antigua could have considered alternatives such as scientific and sustainable tourism or think out of the box by exploring technology and IT development for the export of services and expertise. The type of tourism currently promoted has minimum to no positive impact on the local community and maximum gain for the developers and the central administration. In 2020, we have the luxury of technology that can allow us to do better when it comes to economic development and diversification. Using innovation and ingenuity so much more can be accomplished rather than resorting to old-fashioned unimaginative colonial strategies that have the reputation of Trumpian disasters that ultimately end up in bankruptcy (Eder and Parlapiano 2016).

When taking into account all the reasons why tourist development is a bad idea, they all pale when we consider the fact that strict tourist investment is just bad business and will yield little return in the long term. This is unlike other forms of tourism such as scientific tourism, eco-tourism, and other types of tourism that take into account their long-term effect on the location. All the money spent on tourism can do nothing to reverse the effects of climate change, the local environment, and people (Scott et al. 2012). As sea levels rise and storms intensify, it seems counter-intuitive to choose to aggressively develop in such a vulnerable ecosystem. The majority of construction on Barbuda is happening in places that are at the forefront of protecting the island and are most vulnerable to storms such as Palmetto Point (Figure 7.2). One must ask oneself, why is an investment in an island that was nearly decimated by a hurricane just a few years ago a good idea? How do these investors justify their investment to insurance companies which must be aware of just how vulnerable these investments are to big climate events? By proceeding with this construction, investors are maximizing vulnerability not only on their investment but for all the people who reside on the island.

Finally, it is important to note that the current neocolonial capitalist paradigm is deeply rooted in settler colonialism. Built on the African slave trade and enslavement of Indigenous populations, settler colonialism is not only 'a global and genuinely transnational phenomenon', but continues to persist to this day (Veracini 2010, 2). One of the defining features of settler colonialism is that it is about domesticating and biopolitically managing their respective domestic domains (Veracini 2010). Its second important feature, at least when it comes to the discussion of Barbuda, is the fact that settler colonialism took off because labor was much more profitable in the colonies than it was at home. Through the exploitation of the enslaved people brought from Africa as in the case of Barbuda and local labor in many other places around the globe, the settler colonialists could protect their market at home while exploiting their colonies for cheap raw material which in turn stimulated home production (Horne 2018, 21). This desire by the settlers to avoid imperial interference and to manage the economy, the settlers created 'circumstances where Indigenous and exogenous Others progressively disappear in a variety of ways: extermination, expulsion, incarceration containment, and assimilation for Indigenous peoples (or a combination of all these elements), restriction and selective assimilation for subaltern exogenous Others' (Veracini 2010, 16–17). Although settler colonialism is quite different from modern-day capitalism, many of the events we see in Barbuda showcase some disturbing correlations.

Settler colonialism has not been entirely erased in the modern-day neoliberal world economic order. Aside from millions of people forced into modern-day slavery and human-trafficking, an even greater number are engaged in exploitative labor. Unlike consensual human-trafficking and modern-day slavery, the voluntary nature of exploitative work does not make it any less problematic. Labor practices used by many nations, which sit at the cornerstone of those economies reliant on tourism, capitalize on economic vulnerabilities which force individuals to accept exploitative work. Researchers

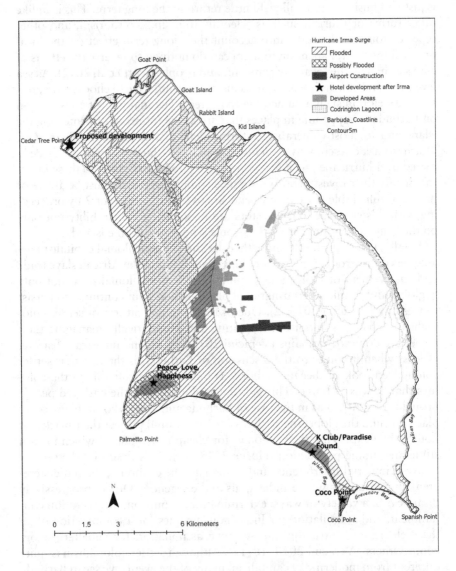

Figure 7.2 Estimates of the extent of the storm surge from Hurricane Irma based on elevation models, National Oceanic & Atmospheric Administration (NOAA) surge estimate of 2.43 m and field observations made in December 2017. All of the tourist developments are well within the surge. While the new airport shown in red is not within the surge, it undergoes seasonal flooding and frequent surface collapses from the caves and tunnels underneath.

argue that 'the political economy of mass tourism is analogous to a master-servant colonial relationship' (Kingsbury 2005). Moreover, mass tourism has created 'plantation-style form of domination that juxtaposes a hedonistic resort-based elite located on the coast with an impoverished, un-skilled, subservient labor-supplying interior' (Kinsbury 2005).

The events that have so far occurred in Barbuda showcase what researchers have identified as part of foreign-owned mass tourism corporation's modus operandi 'loss of control of local resources; low multiplier and spread effects outside of tourism enclaves; lack of articulation with other domestic sectors; and high foreign exchange "leakages"' (Kingsbury 2005). Every single one of these events has occurred in Barbuda, and the tourist resorts have not yet even been opened. Ultimately, tourism is not going to change Barbuda's fortunes, and will only 'reinforce hierarchical relations and skew egalitarian development efforts' (Kingsbury 2005).

7.3 Tangible and intangible cultural heritage and people's livelihood

Numerous governments around the globe have exploited catastrophic events as a means of ensuring a quick profit (Klein 2018; Gunewardena and Schuller 2008). In most cases, this also leads to cutting corners, and where little to no due diligence is paid to how these projects affect the environment. Even less attention is paid to the effects of disaster capitalism on tangible and intangible cultural heritage. Destruction of heritage is problematic (Perdikaris et al. 2017), and while the hurricane has been destructive, the true disaster towards heritage occurs as a result of the bulldozers that come hand in hand with disaster capitalism (Perdikaris et al. 2021b). Ultimately, disaster capitalism leads to the erasure of a culture, resulting in cultural genocide, to be replaced by capitalist-driven white supremacy.

The people of Barbuda have a unique relationship with the communal land which they have occupied for nearly half a millennia. The totality of the landscape and seascape making up Barbuda hold important memories in the understanding of place, tradition, and identity that outsiders might mistakenly identify as just 'uninhabited spaces' (Perdikaris et al. 2021a). Before Irma, Barbuda was home to approximately 1,800 residents, of which only 1,200 have returned, most of whom are descendants of the enslaved people brought to the island in the 1600s. To Barbudans, rocks, caves, and other sites around the island, all have local names, history, and tangible cultural significance. These spaces and sites have a purpose, whether spiritual or practical, and the activities that occur in these spaces have meaning and significance.

On Barbuda, there are numerous tangible heritage sites including historic buildings, museums, monuments, documents, and other artifacts

(Perdikaris 2018; 2020). When these sites are affected, there are visible losses. Intangible cultural heritage is expressed 'through orality, gesture and other forms of expression that individuals create using various media and instruments'. The impacts and long-term effects of the loss or disruption of these traditions are much harder to see (Johannot-Gradis 2015, 1255). In Barbuda, elements of cultural identity such as the large flat rocks outside peoples' homes are not just debris but are an important part of intangible cultural heritage. These stones are gathering places where family and friends come together at the end of the workday to share stories, catch up, and talk politics (L. Thomas, personal communication, January 13, 2014). These rocks are communal gathering places that hold an important cultural meaning in maintaining oral traditions. Whether traditional artistry, festivities, or religious rituals, it is important to recognize that there are various types of cultural heritage, and while to the untrained outsider's eye they seem insignificant, to the people they are an integral part of ethnicity and belonging, and a legacy that is passed onto future generations. These structures and practices ground people in their landscape and create a sense of identity and resilience (Perdikaris et al. 2018, 2021a, 2021b). Maintaining and restoring cultural heritage and traditional lifeways should and must be preserved and not be treated as dispensable by profiteering agendas.

Shortly after Irma in December 2017, two of the authors met with several Barbudans to conduct a visioning exercise. The Barbudans were keenly aware of the strong external forces pushing for the privatization of land and large-scale tourist development that since then has happened. Together, we looked at the island holistically and divided the land into areas based on tangible and intangible cultural heritage, ecological features and sensitivities, and potential for small-scale tourist attractions that would be owned and operated by Barbudans. In this visioning exercise, we followed the principles of Asset-Based Community Development, an approach that builds on the natural and social assets rather than the needs for future development (Haines 2009; Phillips and Pittman 2014). Based on these discussions, we divided Barbuda into eight areas, shown in Figure 7.3, and, then within each area, identified the cultural and natural assets, considerations for the development, and recommendations for the types of tourist development that could bring revenue to Barbudans. For example, Figure 7.4 shows two areas, the Lagoon National Park and Goat Flash area that is within the Ramsar site to the north and the area along the western coast facing the Caribbean sea. We then began identifying the assets, considerations, and recommendations, shown in Table 7.1. These areas have incredible natural and cultural assets that could sustain livelihoods for Barbudans while providing additional revenue for small-scale tourism ventures. The list of items shown in Table 7.1 is not complete but is intended to be a starting point to illustrate the benefits of dialogue among community members that could identify sustainable and resilient paths for Barbuda and the people living there that are not part of the mono-economy of developing tourism.

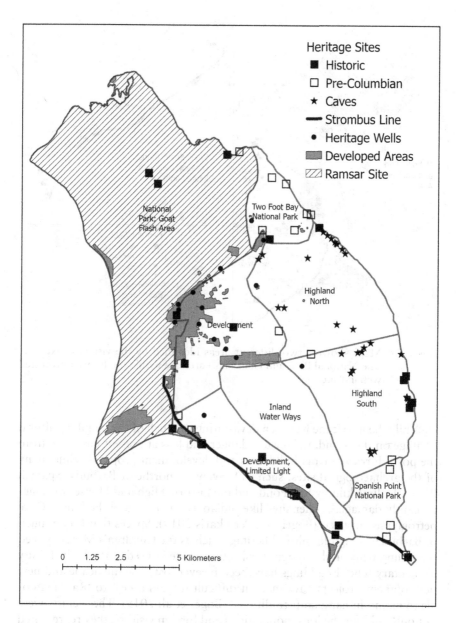

Figure 7.3 Map of Barbudan's vision for eight different areas based on Traditional Ecological Knowledge (TEK), cultural heritage, and understanding of ecology.

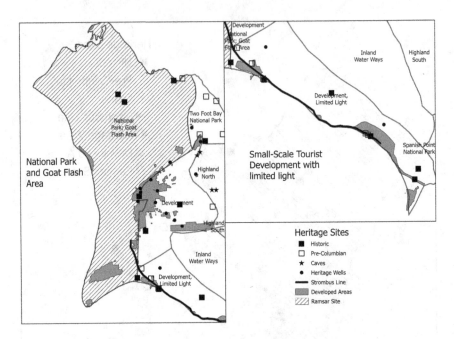

Figure 7.4 Maps showing two different areas identified by the visioning exercise. The National Park and Goat Flash area are within the Ramsar protected wetland site.

While the hurricane has been devastating to much of the tangible cultural heritage on the island, the biggest danger and loss at present has come from the post-Hurricane Irma bulldozers and development projects. While many of the archaeological sites, such as Seaview in northeast Barbuda, Spanish Point, the lime kiln by Salt Pond and the historic Highland House, were significantly damaged, other sites like Indian Town Trail and the Indian Cave petroglyphs survived (Boger and Perdikaris 2019). Spaces that house much of Barbudas' tangible cultural heritage such as the Children's Museum need rebuilding while archaeological collections once housed at the Holy Trinity elementary school buildings have been forever lost. The hurricane and new development projects have made it difficult to gain access to 'Raw materials to maintain handicraft traditions' (Boger et al. 2019). These crafts were not only sold for the local populations and foreign visitors; they represented an important part of how Barbudans maintain their connections to their ancestors and ways of life. Moreover, 'traditions like horse racing, hunting, camping, living in caves, beach festivities, traditional healing, and construction of traditional houses are endangered', all of which puts Barbuda's intangible heritage in danger and on a path to extinction. As one of the local fishermen Devon Warner pointed out in an interview with *The Guardian*

Table 7.1 Visioning exercise suggestions

	National Park and Goat Flash	Western Caribbean sea side
Assets	Existing national park; on list for the UNESCO World Heritage Site Rare and endangered species; frigate bird colony Heritage sites include fishing creek, wells, and pre-Columbian	Extensive beaches and wetlands Sea turtle lay eggs along beaches Small-scale tourist cottages in operation Ferry terminal located here
Considerations	Flooded during Hurricane Irma; will be area inundated early on by sea level rise Strong heritage value for Fishing Creek and Goat Flash Important hunting area for marine and terrestrial species; largest land turtles found here	Turtle hatchlings sensitive to light pollution Subject to storm surges Heritage sites, especially from the Archaic time period Lots of no see ums (bugs)
Recommendations	Reinforce national park management; move forward with UNESCO WHS status Fishing Creek and Goat Flash for Barbudan use Tourism opportunities for birds and other species in national park	Small-scale tourist housing; limit light - use as selling point for ecotourists

The development will have a significant impact on our traditions and cultures which date back as long as I can remember. The way we use our land will change, along with the freedoms we have always enjoyed as parts of Barbuda will be off limits to us.

(*The Guardian* 2020)

Numerous sacred spaces and locations that have provided subsistence or supported traditional activities such as camping, hunting, and gathering, for hundreds of years have been blocked off or destroyed (i.e., the airport area and Palmetto Point). This is especially important when one considers that local plants have more than just ecological and subsistence value; they also hold an important cultural value. From an early age, Barbudans are taught that all one needs when they find themselves in the bush and are hungry are some tamarind and lime (J. Mussington, personal communication, December 3, 2020). This type of knowledge isn't written, but learned and passed from generation to generation as a critical oral tradition. Until 2018, the land was communally owned, but now has been designated for private development. Even though the international laws regarding the shoreline and the now repealed Barbuda Land Act make it illegal for the development projects to keep people out, they have nevertheless done so. The Codrington Lagoon and surrounding areas (see Figure 7.1) are within a designated Ramsar site for protected wetlands. Developing within the Ramsar site puts these ecosystems at risk, is in violation of requirements for the Ramsar biodiversity designation, and also could be in violation of international obligations for the Convention on Biological Diversity (Phillips 2020).

The questionable activities by the Antiguan government have thus significantly affected the Barbudan way of life, and their ability to be resilient and survive sustainably on their island. Barbudans no longer have access to resources, including land (such as specific wells and areas for hunting), or spaces where small-scale agriculture was practiced. For hundreds of years, Barbudans relied on crop and plot rotation to ensure that enough nutrients accumulate in the soil to support swidden agriculture. Seemingly vacant space is a space of purpose. The low fertility land in Barbuda can be productive and sustain the local resident population when cultivated in this traditional rotation pattern. The caves that have provided not only shelter and safety for Barbudans for hundreds of years also hold spiritual and cultural meaning (Boger et al. 2016; Perdikaris et al. 2013). Development projects run the risk of affecting agriculture, making spiritual and cultural locations inaccessible, along with the environmental effects such as polluting the island's aquifer (through the discard of desalination byproducts along with chemicals used for the proposed golf courses) without which Barbudans will no longer have access to free, clean, and safe water (Boger et.al. 2014, 333).

Beyond the destruction of the tangible and intangible cultural heritage and its effects on Barbudan culture, there is also a question about the ability of the ecosystem services to support such an increase in population and infrastructure. The population increase that is inferred by these development activities will bring challenges to the long-term survival of the island as a whole. Energy, waste, potable water, food production, and safety on a karstic terrain, along with resiliency under the constraints of big climatic events in the region, should be a cause for concern. Building resorts and increased housing for foreign workers and tourists creates a big demand on all levels and pushes the island further away from its traditional sustainable existence.

7.4 Environment and disaster capitalism

Barbudans have a long history of resilience and adaptation; their special relationship to the environment has allowed them to survive through droughts, increased rainfall, and hurricanes (Bain et al. 2017; Burn et al. 2015). The communally owned land has allowed Barbudans to remain one of the 'last Caribbean islands that practices hunting and gathering and living from the land' (Boger et al. 2014, 328). The Barbudans' focus on subsistence agriculture, fishing, charcoal, and open-range livestock (Berleant-Schiller 1977; Lowenthal and Clarke 2007) has allowed them to live sustainably for over 400 years. Barbudan people are an important component of the ecosystem makeup of the island; Barbudan culture, flora, and fauna coexist and co-evolve. While flora and fauna may have been different before the colonial period, the present biodiversity includes many rare and endangered species that continue to be protected by the Barbudan way of life.

The newest developments have attacked some of Barbuda's most precious natural resources, its wetland sites, lagoon, dunes, and all the biodiversity included within them. These locations were all once protected under the 2007 Barbuda Land Act. The lagoon and surrounding areas were deemed of international importance and were declared protected under the Ramsar bio-diversity allocation in the Geneva-based Convention on Wetlands (Ramsar Secretariat) (see Figure 7.1), of which Antigua and Barbuda are signatories (George 2020). As the signatories of the convention, the government of Antigua and Barbuda is obligated to protect wetlands and must adhere to wetland conservation in its national land-use planning (see Figure 7.2).

The location where Barbuda's only protected wetland existed is now home to a construction project, Peace, Love, and Happiness (see Figure 7.2), that will see a 'construction of a marina within the lagoon, 81 beach front lots, high-end cottages and homes as well as a golf course with about 400 homes on 650 acres of land' (George 2020). The wetland is Category II Protected Area under the International Union for Conservation of Nature and a designated Wetland of International Importance under the Ramsar Convention, site no. 1488 recognized since 2005 (GLAN 2020). The destruction of the wetland will have permanent detrimental effects on the island's ecosystem. According to local marine biologist John Mussington

> The West Indian whistling duck – which is critically endangered – is dependent on areas like this to reproduce. Five thousand magnificent frigate birds come to our lagoon each nesting season. And for the endemic Barbuda warbler, this place is critical to its survival. These wetlands are crucial to our coral reef health and our marine resources.
>
> (*The Guardian* 2020)

On September 12, 2020, the Global Legal Action Network submitted a complaint to the Ramsar Secretariat requesting the investigation of the destruction of wetlands in the Codrington Lagoon National Park (George 2020). As of February 2021, despite court orders on October 1, 2020, to halt construction, locals have witnessed ongoing operations and two members of the Barbuda Council were arrested while trying to inspect the site (GLAN 2020).

The Codrington Lagoon is home to a very successful lobster nursery whose genetic footprint can be found throughout the Caribbean (Ruttenberg et al 2018). The lobster fishery export provides an important economic revenue for locals, and it is one of the few sustainable sources of income on the island. The construction of the proposed marina, golf course, and high-end luxury homes for the Peace, Love, and Happiness venture has led to deforestation and sand mobilization, and ultimately will degrade the lagoon's ecology (George 2020). The development has thus placed the whole ecosystem of the lagoon in danger. A small limestone island with thin soils is a landscape and seascape that needs to be carefully managed.

Ten years ago, the Barbuda Council voted against a proposal to intensify the production of poultry and pigs (J. Mussington, personal communication, January 7, 2015). This decision was not made to discourage individual production, but the Council acknowledged that production on an industrial scale would contaminate and affect the ecological balance. An increased number of humans on the island and poor planning will have the same devastating environmental effect. Artificial barriers, destruction of mangroves, increased food and water demand, and excessive solid and liquid waste will put unprecedented pressure on the island resources and its ability to maintain healthy, resilient ecosystems. Habitats for endemic species, economic species, and rare/protected reefs and vegetation will be threatened. Barbuda's unspoiled beaches have survived the passage of time and are host to nesting sea turtles; noise and light pollution, along with human presence, are minimal. The construction of two luxury residences will begin along beaches that are home to 'well-known nesting habitats for endangered and threatened sea turtles and other wildlife creatures' (George 2020). Undoubtedly, these ecosystems will be severely impacted while rare and endangered species will be further threatened.

Development projects are placing the entire island and its population in danger by destroying dune systems in the Palmetto Point area. Dune systems protect the land during storms and wave surges while also providing filtration to rain and salt water that filters through the sand into the aquifer. The Palmetto aquifer feeds into some of the island's historic wells that locals are still actively using for their everyday needs (Boger et al. 2014). It took 5,000 years for the sand dunes to develop (Brasier and Donahue 1985). Barbuda's ecological survival is at risk. Deals with the central government in Antigua allow developers to ship containers straight to Barbuda and bypass all import regulations and customs. While these practices are not technically illegal, they are nevertheless an indicator of how little the government in Antigua cares about following proper import channels that are designed to protect local ecosystems. In addition to revenue, custom checkpoints are important ways to control the unintended import of pests and disease, which is particularly important for island ecosystems that may have no natural defenses to these external stressors. Antigua, for instance, is plagued by the Giant African Snail, which has compromised food security and the agricultural sector on the island while non-native beetles are destroying palm trees and coconut trees. These are a few of the long list of invasive species (The Daily Observer 2019; Intra-American Institute for Cooperation in Agriculture 2019).

Construction on the island failed to use expert opinion before beginning construction, best exemplified in the post-Irma airport. Given the limestone geology, the island is full of sinkholes, caves, and underground tunnels. This type of geology is problematic for a large-scale airport; an environmental assessment would have quickly revealed this fact. However, construction began at a rapid pace, 24 hours for 7 days a week, and without surprise, cave-ins quickly followed. The construction company only stopped after multiple sinkholes and cave-ins. The international airport project is currently held in litigation, and construction has halted. As we write this chapter, a public document is circulating calling for public comments about the environmental study done by the investment company claiming that the construction can continue. One of the elements in the document is the use of a map from Boger et al. (2014), claiming that there are no archaeological sites along the new development sites because there are none listed in that publication. Clearly, articles written for scientific research are not meant to be a survey or study or a specific location. None of the authors of the 2014 article have ever been contacted by this firm to provide a proper assessment of the sites in question. The spokesperson for PLH (the primary developer on the island) claims that 'Our environmental impact study was done by an established, internationally recognized scientist. We are yet to see credible science that contradicts what it has outlined' (*The Guardian* 2020). The environmental impact study misrepresented information from previous studies. As a result of this, species unique to Palmetto have been destroyed and cannot be replanted. The Palmetto Point is not only significant environmentally, but is home to 'culturally vital vegetation (broom making, thatch roofs, fish nets, medicinal plants)' (Boger et al. 2014, 331). Additionally, development in this area increases exposure to hurricanes.

It's important to note that the Ramsar Convention is not the only international agreement that the Antiguan government has signed and then failed to uphold. As the destruction of Barbuda continues, the Antiguan government has ratified the Escazú Agreement. The Escazú Agreement is short for Regional Agreement on Access to Information, Public Participation, and Justice in Environmental Matters in Latin America and the Caribbean adopted on March 4, 2018, and was opened for signatures at the United Nations Headquarters in New York from September 27, 2018, to September 26, 2020, by any of the countries of Latin America and the Caribbean. Antigua and Barbuda ratified the agreement on March 4, 2020, (United Nations Treaty Collection). The Escazú Agreement is a pact between Latin American and Caribbean countries seeking to reduce social conflicts and protect frontline communities in the world's deadliest region for environmental defenders (Bermúdez Liévano 2020). In addition to protecting the environmental defenders, the agreement also calls for 'communities right to consultation on the impacts of large development projects' (Bermúdez Liévano 2020). The irony of this agreement cannot be lost as the same government that has spearheaded this agreement has also done nothing to ensure that the agreement is observed in its own country. Without any consultation with the local communities, the Antiguan government has done exactly what they promised not to do, and that is to disregard peoples' opinions regarding development projects on their land.

7.5 Conclusion

Barbuda finds itself at an important precipice where important decisions about the future must be made. Barbudans at the moment are divided between some who have embraced the development and others who seek the support of foreign organizations to help them protect what is left. The future of Barbuda hangs in the balance, and questions such as equity, climate change, inclusion, and survival are at the core of these decisions.

With roots in the exploitative practices of settler colonialism, disaster capitalism remains a dominant force in modern times. While natural disasters are responsible for the initial disruptions to the ways of life in the regions where they occur, disaster capitalism often creates an even greater chasm and more permanent damage. This process is not only dangerous but unsustainable. It is estimated that although Indigenous peoples comprise less than 5% of the global population, they protect 80% of global diversity (Raygorodetsky 2018). While Barbudans may not be called Indigenous, they share many characteristics, with Indigenous peoples, especially their land-based identities (Perdikaris et al., 2021a), and as such are part of a small group of people protecting global resources for which the rest of the world relies on. As demonstrated in the visioning exercise, an alternate, more resilient, and sustainable path for Barbuda and Barbudans that can bypass the pitfalls of 20th-century large-scale tourism development is within reach. Combining traditional ecological knowledge with in-depth environmental assessments conducted by objective scientists will create the right path to a resilient future that will more effectively balance economic diversification and ecological well-being while maintaining the richness of Barbudan cultural identity.

References

Bader, C., Bieri, S., Wiesmann, U. and Heinimann, A., 2017. Is economic growth increasing disparities? A multidimensional analysis of poverty in the Lao PDR between 2003 and 2013. *The Journal of Development Studies*, 53(12), pp. 2067–2085.

Bain, A., Faucher, A. M., Kennedy, L., LeBlanc, A. R., Burn, M., Boger, R., Perdikaris, S., 2017. Early landscape transformations of Barbuda, West Indies. New Research in Environmental Archaeology and Palaeoecology. Environmental Archaeology: The Journal of Human Palaeoecology (Wallman & Rivera guest editors).pp. 36–46.

Berleant-Schiller, R., 1977. The social and economic role of cattle in Barbuda. *Geographical Review*, 67, pp. 299–309.

Bermúdez Liévano, A., 2020. Diálogo Chino. Escazú Agreement gains momentum. March 17, 2020. https://dialogochino.net/en/infrastructure/34120-the-escazu -agreement-gathers-half-of-the-countries-needed-to-become-a-reality/

Birau, R. and Doaga, D.I., 2019. The effects of poverty on emerging countries development in the context of globalization and rapid economic growth. *Annals of the Constantin Brancusi University of Targu Jiu-Letters & Social Sciences Series*, 1/2019, pp. 107–114.

Boger, R. and Perdikaris, S., 2019. After Irma, disaster capitalism threatens cultural heritage in Barbuda. *NACLA*, February, *11*, p. 2019.

Boger, R., Perdikaris, S., Potter, A.E. and Adams, J., 2016. Sustainable resilience in Barbuda: Learning from the past and developing strategies for the future. *International Journal of Environmental Sustainability*, *12*(4), pp. 1–14.

Boger, R., Perdikaris, S., Potter, A.E., Mussington, J., Murphy, R., Thomas, L., Gore, C. and Finch, D., 2014. Water resources and the historic wells of Barbuda: Tradition, heritage and hope for a sustainable future. *Island Studies Journal*, *9*(2), pp. 327–342.

Boger, R., Perdikaris, S. and Rivera-Collazo, I., 2019. Cultural heritage and local ecological knowledge under threat: Two Caribbean examples from Barbuda and Puerto Rico. *Journal of Anthropology and Archaeology*, *7*(2), pp. 1–14.

Bourne, C., 2009. Economic growth, poverty and income inequality. *Journal of Eastern Caribbean Studies*, *34*(4), pp. 21–48.

Brasier, M. and Donahue, J., 1985. Barbuda—An emerging reef and lagoon complex on the edge of the Lesser Antilles island are. *Journal of the Geological Society*, *142*(6), pp. 1101–1117.

Brosnan, D. and Associates., 2020. Low-density private residential development on cedar tree point: Environmental impact assessment. November 4, 2020. https:// environment.gov.ag/assets/uploads/attachments/a4fbc-cedar-tree-point-eia-11.4 .20_reduced-file-size.pdf.

Burn, M.J., Holmes, J., Kennedy, L. M., Bain, A., Marshall, J.D. and Perdikaris, S. 2015. A sediment-based reconstruction of Caribbean effective precipitation during the "little Ice Age" from freshwater pond, Barbuda. The Holocene I-II Sage publications. Sagepub.co.uk.

Chappell, B., 2017, September 29. 3 weeks after Irma Wrecked Barbuda, island lifts mandatory evacuation order. *NPR*. https://www.npr.org/sections/thetwo -way/2017/09/29/554540066/3-weeks-after-irma-wrecked-barbuda-island-lifts -mandatory-evacuation-order.

Coats, A.W., 1979. Adam Smith and the wealth of nations, 1776–1976: Bicentennial essays. *The Journal of Economic History* 39, 3 pp 836–838.

Eder, S. and Parlapiano, A., 2016. Trump's projects: Comparing hype with results. (cover story). *New York Times*, October 7. http://search.ebscohost.com .libproxy.unl.edu/login.aspx?direct=true&db=aph&AN=118606561&site =ehost-live.

Edward, P. and Sumner, A., 2018. Global poverty and inequality: Are the revised estimates open to an alternative interpretation? *Third World Quarterly*, *39*(3), pp. 487–509.

George, E., 2020. Int'l lawyers request investigation of wetland destruction on Barbuda. *The Daily Observer*, December 10, 2020. Accessed December 13, 2020. https://antiguaobserver.com/intl-lawyers-request-investigation-of-wetland- destruction-on-barbuda/?fbclid=IwAR3DxyzXtk07zc1uiQ7SXHD -QZG6aH0iqL9VQqOf_SzEhcez4F6x9GBxaRc.

Global Legal Action Network (GLAN), 2020. Land grab & wetland destruction in Barbuda. Glanlaw.org. https://www.glanlaw.org/barbudalandgrab

Gould, K.A. and Lewis, T.L., 2018. Green gentrification and disaster capitalism in Barbuda. *NACLA Report on the Americas*, *50*(2), pp. 148–153.

Gunewardena, N. and Schuller, M., 2008. *Capitalizing on Catastrophe: Neoliberal Strategies in Disaster Reconstruction*. Lanham, MD: AltaMira Press.

Haines, A., 2009. Asset-based community development. In R. Phillips and R.H. Pittman (Eds.) *An introduction to community development*, London: Routledge Taylor and Francis Group, *38*, p. 48.

Horne, G., 2018. The apocalypse of settler colonialism: The roots of slavery, white supremacy, and capitalism in 17th century North America and the Caribbean, monthly review press. ProQuest Ebook Central. https://ebookcentral-proquest-com .libproxy.unl.edu/lib/unebraskalincoln-ebooks/detail.action?docID=4844848.

Inter-American Institute for Cooperation in Agriculture, 2019. Communities in Antigua and Barbuda advance in the control of the giant African snail with the support of IICA. https://www.iica.int/en/press/news/communities-antigua-and -barbuda-advance-control-giant-african-snail-support-iica.

Johannot-Gradis, C., 2015. Protecting the past for the future: How does law protect tangible and intangible cultural heritage in armed conflict?. *Int'l Rev. Red Cross*, *97*, p. 1253.

Kingsbury, P., 2005. Jamaican tourism and the politics of enjoyment. *Geoforum*, *36*(1), pp. 113–132.

Klein, N., 2018. *The battle for paradise: Puerto Rico takes on the disaster capitalists*. Chicago: Haymarket Books.

Lowenthal, D. and Clarke, C., 2007. The Triumph of the commons: Barbuda belongs to all Barbudans together. In J. Bensson and J. Momsen (Eds.) *Caribbean land and development revisited*. New York: Palgrave Macmillan, pp. 147–158.

Manzetti, L., 2009. *Neoliberalism, accountability, and reform failures in emerging markets: Eastern Europe, Russia, Argentina, and Chile in comparative perspective*. University Park: Penn State University Press.

Perdikaris, S., 2018. The sea will rise, barbuda will survive. In *Environmental justice as a civil right*. La Bienalle di Venezia. 16. Mostra Internazionale di Architectura, pp. 56–58.

Perdikaris, S., 2020. What is environmental consciousness? A thematic cluster. *Ecocene: Cappadocia Journal of Environmental Humanities*, *1*(2), 1–4. https:// doi.org/10.46863/ecocene.2020.17

Perdikaris, S., Bain, A., Boger, R., Grouard, S., Faucher, A.M., Rousseau, V. and Medina-Triana, J., 2017. Cultural heritage under threat: The effects of climate change on the small island of Barbuda, Lesser Antilles. In: T. Dawson, C. Nimura and E. Lopez-Romero (Eds.) *Public Archaeology and Climate Change*. Oxford: Oxbow Books Chapter 15, pp. 138–148.

Perdikaris, S., Bain, A., Grouard, S., Baker, K., Gonzalez, E., Hoelzel, A.R., Miller, H., Persaud, R. and Sykes, N., 2018. From icon of empire to national emblem: new evidence for the fallow deer of Barbuda. *Environmental Archaeology*, *23*(1), pp. 47–55.

Perdikaris, S., Boger, R., Gonzalez, E., Ibrahimpašić, E. and Adams, J.D., 2021a. Disrupted identities and forced nomads: A post-disaster legacy of neocolonialism in the island of Barbuda, Lesser Antilles. Nomadic identities. *Island Studies Journal*, *16*(11), pp. 115–134. Edited by May Joseph. Editor in Chief Adam Grydehoj.

Perdikaris, S., Boger, R. Ibrahimpašić, E., 2021b. Seduction, promises and the Disneyfication of Barbuda post Irma. *Translocal Contemporary Local and Urban Cultures Journal, 5*(2): pp. 1–15. Number 5 (un)inhabited spaces. Ana Salgueiro and Nuno Marques (eds). ISSN 2184-1519 Madeira and Umeå. Translocal.cm-f unchal.pt/2019/05/02/revista05/.

Perdikaris, S., Grouard, S., Hambrecht, G., Hicks, M., Mebane Cruz, A. and Peraud, R., 2013. The caves of Barbuda's eastern coast: Long term occupation, ethnohistory and ritual. *Caribbean Connections, 3*, pp. 1–9.

Perdikaris, S., Hejtmanek, K., Boger, R., Adams, J.D., Potter, A.E. and Mussington, J., 2013. The tools and technologies of transdisciplinary climate change research and community empowerment in Barbuda. *Anthropology News, 54*(2). https://anthrosource.onlinelibrary.wiley.com/doi/abs/10.1111/j.1556-3502.2013.54202.x

Perdikaris, S. and Hejtmanek, K.R., 2020. The sea will rise, Barbuda will survive: environment and time consciousness. *Ecocene: Cappadocia Journal of Environmental Humanities, 1*(2), pp. 92–108.

Phillips, R. and Pittman, R. eds., 2014. *An introduction to community development.* London: Routledge.

Phillips, Z.A.R., 2020. Barbuda's community title to land: A furtherance of the convention on biological diversity? *Review of European, Comparative & International Environmental Law, 29*(1), pp. 118–128.

Purvis, B., Mao, Y. and Robinson, D., 2019. Three pillars of sustainability: In search of conceptual origins. *Sustainability Science, 14*, pp. 681–695. https://doi.org/10.1007/s11625-018-0627-5

Raygorodetsky, G., 2018. Indigenous peoples defend earth's biodiversity–But they're in danger. *National Geographic.* https://www.nationalgeographic.com/environment/article/can-indigenous-land-stewardship-protect-biodiversity

Regional agreement on access to information, public participation and justice in environmental matters in Latin America and the Caribbean. Escazú, 4 March 2018, United Nations Treaty Collection, available from https://treaties.un.org/pages/ViewDetails.aspx?src=TREATY&mtdsg_no=XXVII-18&chapter=27&clang=_en.

Ruttenberg, B., Caselle, J.E., Estep, A.J., Johnson, A.E., Marhaver, K.L., Richter, L.J., Sandin, S.A., Vermeij, M.J., Smith, J.E., Grenda, D. and Cannon, A., 2018. Ecological assessment of the marine ecosystems of Barbuda, West Indies: Using rapid scientific assessment to inform ocean zoning and fisheries management. *PloS One, 13*(1), p. e0189355.

Schuller, M. and Maldonado, J.K., 2016. Disaster capitalism. *Annals of Anthropological Practice, 40*(1), pp. 61–72.

Scott, D., Gössling, S. and Hall, C.M., 2012. International tourism and climate change. *Wiley Interdisciplinary Reviews: Climate Change, 3*(3), pp. 213–232.

Stonich, S., 2008. International tourism and disaster capitalism. In N. Gunewardena and M. Schuller, (Eds.) *Capitalizing on catastrophe: Neoliberal strategies in disaster reconstruction.* Maryland: Roman and Littlefield Publishers, Inc. Altamira Press. pp. 47–68.

The Daily Observer., 2019. Community effort needed to eradicate the giant African Snail. June 5, 2019. https://antiguaobserver.com/community-effort-needed-to-eradicate-the-giant-african-snail/.

The Guardian., 2020. Barbudans 'fight for survival' as resort project threatens islanders' way of life. *The Guardian*, December 4 [Online]. Accessed December 14, 2020. https://www.theguardian.com/world/2020/dec/14/barbuda-luxury-resort-project-dispute?fbclid=IwAR12GPGR-2kA7eZK0zfAcF3X3KrpFDwXqm0U pZfkTaU-0wFEvmzOzlu7bRk.

Veracini, L., 2010. *Settler colonialism: A theoretical overview*. London: Palgrave Macmillan UK. ProQuest Ebook Central.

Yee, D.K.P., 2018. Violence and disaster capitalism in post-Haiyan Philippines. *Peace Review*, 30(2), pp. 160–167.

Index

Printed in the United States
by Baker & Taylor Publisher Services

Printed in the United States
by Baker & Taylor Publisher Services